VisualDSolve

Dan Schwalbe   Stan Wagon

VisualDSolve

Visualizing Differential Equations
with *Mathematica*®

Dan Schwalbe
Stan Wagon
Department of Mathematics and
    Computer Science
Macalester College
St. Paul, MN 55105, USA

Publisher: Allan M. Wylde
Publishing Associate: Keisha Sherbecoe
Product Manager: Walter Borden
Production Manager: Natalie Johnson
Manufacturing Supervisor: Jacqui Ashri
Cover Designer: Trina Baucomb

Library of Congress Cataloging-in-Publication Data
Schwalbe, Dan
    VisualDSolve:Visualizing Differential Equations with Mathematica/Dan
    Schwalbe, Stan Wagon.
        p. cm.
    ISBN 978-0-387-94721-1 (sftcvr: alk. paper)
        1. VisualDSolve.   2. Differential equations—Numerical solutions—
    Date processing.   I. Wagon, S.   II. Title.
    QA371.5.D37S39 1996
    515´.352´0113668—DC20   96-42076

Printed on acid-free paper.

ISBN-13: 978-1-4612-7473-5   e-ISBN-13: 978-1-4612-2251-4
DOI: 10.1007/978-1-4612-2251-4

Photocomposed pages prepared from the authors' *Mathematica* files.

9 8 7 6 5 4 3 2 1

Additional material to this book can be downloaded from http://extras.springer.com.

ISBN-13: 978-1-4612-7473-5   SPIN 10532318

THE
ELECTRONIC
® LIBRARY
OF
SCIENCE

TELOS, The Electronic Library of Science, is an imprint of Springer-Verlag New York with publishing facilities in Santa Clara, California. Its publishing program encompasses the natural and physical sciences, computer science, mathematics, economics, and engineering. All TELOS publications have a computational orientation to them, as TELOS' primary publishing strategy is to wed the traditional print medium with the emerging new electronic media in order to provide the reader with a truly interactive multimedia information environment. To achieve this, every TELOS publication delivered on paper has an associated electronic component. This can take the form of book/diskette combinations, book/CD-ROM packages, books delivered via networks, electronic journals, newsletters, plus a multitude of other exciting possibilities. Since TELOS is not committed to any one technology, any delivery medium can be considered. We also do not foresee the imminent demise of the paper book, or journal, as we know them. Instead we believe paper and electronic media can coexist side-by-side, since both offer valuable means by which to convey information to consumers.

The range of TELOS publications extends from research level reference works to textbook materials for the higher education audience, practical handbooks for working professionals, and broadly accessible science, computer science, and high technology general interest publications. Many TELOS publications are interdisciplinary in nature, and most are targeted for the individual buyer, which dictates that TELOS publications be affordably priced.

Of the numerous definitions of the Greek word "telos," the one most representative of our publishing philosophy is "to turn," or "turning point." We perceive the establishment of the TELOS publishing program to be a significant step forward towards attaining a new plateau of high quality information packaging and dissemination in the interactive learning environment of the future. TELOS welcomes you to join us in the exploration and development of this exciting frontier as a reader and user, an author, editor, consultant, strategic partner, or in whatever other capacity one might imagine.

TELOS, The Electronic Library of Science
Springer-Verlag Publishers
3600 Pruneridge Avenue, Suite 200
Santa Clara, CA 95051

THE
ELECTRONIC
LIBRARY
OF
SCIENCE

TELOS Diskettes

Unless otherwise designated, computer diskettes packaged with TELOS publications are 3.5" high-density DOS-formatted diskettes. They may be read by any IBM-compatible computer running DOS or Windows. They may also be read by computers running NEXTSTEP, by most UNIX machines, and by Macintosh computers using a file exchange utility.

In those cases where the diskettes require the availability of specific software programs in order to run them, or to take full advantage of their capabilities, then the specific requirements regarding these software packages will be indicated.

TELOS CD-ROM Discs

For buyers of TELOS publications containing CD-ROM discs, or in those cases where the product is a stand-alone CD-ROM, it is always indicated on which specific platform, or platforms, the disc is designed to run. For example, Macintosh only; Windows only; cross-platform, and so forth.

TELOSpub.com (Online)

Interact with TELOS online via the Internet by setting your World-Wide-Web browser to the URL: *http://www.telospub.com*.

The TELOS Web site features new product information and updates, an online catalog and ordering, samples from our publications, information about TELOS, data-files related to and enhancements of our products, and a broad selection of other unique features. Presented in hypertext format with rich graphics, it's your best way to discover what's new at TELOS.

TELOS also maintains these additional Internet resources:

gopher://gopher.telospub.com
ftp://ftp.telospub.com

For up-to-date information regarding TELOS online services, send the one-line e-mail message:

send info to: *info@telospub.com*.

Dedicated to the memory of Jerry Keiper, a superb numerical analyst and an inspirational teacher who was killed in a tragic bicycle accident in Champaign, Illinois, in January 1995.

Preface

This book is a guide to `VisualDSolve`, a comprehensive *Mathematica* package that provides a wide variety of tools for the visualization of solutions to ordinary differential equations. While part of what the package does is to implement standard ideas, such as the generation of the graphs of solutions and the orbits corresponding to solutions of systems, we have also included some new ideas in the visualization of differential equations. Primary among these is the use of shaded gray regions in the phase plane, where the shading is according to the four possible directions of the underlying vector field. This is accomplished via the `NullclineShading` option, presented in section 4.7. Another unusual idea is the use of curvy fish shapes to represent the flow of a vector field (the `FlowField` option, section 4.3).

Many of our ideas are possible only because of *Mathematica*'s great power and openness. In particular, it is possible to access the data in a contour plot. This can be used to great advantage and underlies our nullcline shading and equilibrium-point finding algorithms.

The five chapters of Part 1 form a manual for the dozens of functions and options in the package. Here is a quick overview. For a single ordinary differential equation for which the user wants to see direction fields and/or plots of solutions, perhaps for several initial values, the `VisualDSolve` function should be used. For a system of two or more differential equations, the user must decide whether he or she wishes to see graphs of the individual functions against time (x vs. t, y vs. t), or the orbits in the phase plane (x vs. y) or phase space (x vs. y vs. z). For graphs, use `SystemSolutionPlot`; for phase space views, use `PhasePlot`. Finally, `SecondOrderPlot` can be used on a second-order equation or system. It translates the input to a first-order system and calls the appropriate function (`SystemSolutionPlot` or `PhasePlot`) according to the exact syntax of the `SecondOrderPlot` call. Direction fields, whether arrows or fish shapes, nullclines, equilibria points, and eigenvalue information are all obtained by setting various options to `PhasePlot` or `VisualDSolve`.

In addition, there are some auxiliary functions that perform some specialized tasks. These are described in Chapter 2.

Part 2 consists of twelve chapters that illustrate the many ways one can use the package to better understand solutions to differential equations. Chapters 6 and 7 are lab notebooks, meant to be used in an interactive way. Several exercises are included, and instructors should consider assigning these labs so that students can familiarize themselves with the basics of *Mathematica* and with the VisualDSolve package (the accompanying disk contains all the outputs in Chapters 6 and 7; the instructor will want to delete them). Chapters 8 through 17 present various aspects of modeling. Some of these chapters discuss material that will be covered in all differential equations courses (such as linear systems and Hamiltonian systems), while others present self-contained modeling projects (such as the flow of lead in a human, the swings of a double pendulum, or the flight of a discus). One point we emphasize throughout is the necessity of learning how to double-check the results of a numerical algorithm for accuracy. This point is particularly addressed in Chapters 10, 11, and 15.

The material we present covers many of the topics generally emphasized in a first course in differential equations. In our course, we emphasize numerical computation of solutions and examination of many models. This seems to be the trend in the teaching of differential equations, and so we feel that our package and book can be profitably used us a supplement, in either an introductory differential equations course or a modeling course.

We are grateful to Steve Izen (Case Western Reserve University) and Ed Packel (Lake Forest College) for several helpful suggestions regarding features to implement and bugs to remove. Jim Doyle (Macalester) helped us with various physics-related issues. Rob Knapp and Jerry Keiper (Wolfram Research, Inc.) provided invaluable help in understanding the inner workings of NDSolve. Knapp and other members of the WRI staff made very important suggestions regarding package design. And we are also indebted to the workbook written by Borrelli, Coleman, and Boyce [BCB], which is a splendid source of interesting examples. Finally, we are very grateful to Paul Wellin, our editor at TELOS, for his constant support and help with all aspects of this project.

If readers come up with interesting ways to use the VisualDSolve package, or have suggestions for future versions of the package, we would be very pleased to hear about them.

Dan Schwalbe (schwalbe@macalester.edu)

Stan Wagon (wagon@macalester.edu)

August, 1996

WEB NOTES. Some material related to the VisualDSolve project (sample chapters, a graphics gallery, animations, update information) is available on the World Wide Web at http://www.telospub.com.

Contents

PART

1

The VisualDSolve Manual

Chapter 1

VisualDSolve

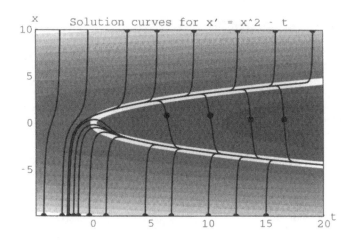

Overview

The `VisualDSolve` command plots solutions to a single first-order ordinary differential equation. The user can specify any number of initial conditions, and options are provided to control the style of the solution curves. Important options include the addition of direction fields, isocline lines, and shaded isocline regions. The method of solution can be symbolic or numeric.

FUNCTIONS DISCUSSED: `DirectionFieldPlot`, `LastPoint`, `Solution`, `VisualDSolve`, `WindowExitData`.

IMPORTANT SUGGESTION. This chapter discusses many features that will be common to all chapters, so the reader should read through the examples here, so as to get a good feeling for the general features of the package.

1.1 ■ Loading the Package

Before any of the commands in the `VisualDSolve` package can be used, the package itself must be loaded. There are several ways of doing this. Two of them are as follows. The main difference is that `Needs` does not do anything if the package has already been loaded; `Get` loads the package whether or not is has already been loaded.

```
<< VisualDSolve`    (* same as Get["VisualDSolve`"] *)

Needs["VisualDSolve`"]
```

There are several ways to tell whether a package has been loaded. The simplest method is to ask for information about one of the functions; if a usage message is returned, then the loading was successful.

```
?VisualDSolve
```

VisualDSolve[equation, {t, tmin, tmax}, {x, xmin, xmax}, (opts)]
   creates an image that shows the solutions to a first-order ODE
   superimposed on a field of slope lines. There are LOTS of
   options! And all of the options for Graphics, Plot, ContourPlot,
   and NDSolve may be used as well. The equation must be input in
   the standard form such as x'[t] + x[t] == t + x[t]^2. Note also
   that the first iterator defines the independent variable and the
   other iterator defines the dependent variable.

Another way is to check the context path via the system variable $ContextPath; if it includes the context of the package in question, then the package has been loaded.

```
$ContextPath
```

{VisualDSolve`, Graphics`Colors`, Utilities`FilterOptions`,
   Global`, System`}

Note that the Colors context is listed. That is because the VisualDSolve package calls the Graphics`Colors` package, thus allowing color names such as Red or DodgerBlue or PeachPuff to be used. To see all the colors in this package, we could use a wild card as follows.

```
?Graphics`Colors`*
```

Alternatively, AllColors returns the complete list. Short[..., 8] restricts the output of expr to 8 lines.

```
Short[AllColors, 8]
```

{AliceBlue, AlizarinCrimson, Antique, Aquamarine,
   AquamarineMedium, AureolineYellow, Azure, Banana,
   Beige, Bisque, Black, BlanchedAlmond, Blue,
   BlueLight, BlueMedium, BlueViolet, Brick, Brown,
   BrownOadder, <<165>>, ViridianLight, WarmGray,
   Wheat, White, Yellow, YellowBrown, YellowGreen,
   YellowLight, YellowOchre, Zinc}

A Common Error

If the user attempts to use one of the package's functions without having first loaded the package, several complications ensue. For example, if the package has not been loaded, an attempt to execute `PhasePlot[...]` creates a "global" variable called `PhasePlot`, which will cast a shadow over the variable of the same name in the package. Short of rebooting, this sort of mistake can be erased via, for example, `Remove[PhasePlot]`. To be more precise, we are seeing here a name conflict between two variables whose full names are `Global`PhasePlot` and `VisualDSolve`PhasePlot`; the prefixes are called *contexts*. A variable such as `x` created by the user is given the `Global`` context; thus its full name is really `Global`x`. But names within a package are given a context related to the name of the package.

1.2 ▪ Basic Usage

To see the solutions to the fundamental equation $dx/dt = x(t)$, we have to specify the equation, the limits on t and x, and some initial values. This last is done via the `InitialValues` option; one can insert specific initial values or simply ask for an $n \times n$ grid of initial values by using `Grid[n]` as a setting. `Needs` loads the package if that has not been done.

```
Needs["VisualDSolve`"];
```

```
VisualDSolve[x'[t] == x[t], {t, -1, 1}, {x, -2, 3},
    InitialValues -> Grid[4]];
```

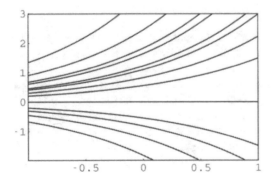

Here's another example.

```
VisualDSolve[x'[t] == Sin[t] x[t], {t, 0, 8}, {x, 0, 3},
    InitialValues -> Grid[4]];
```

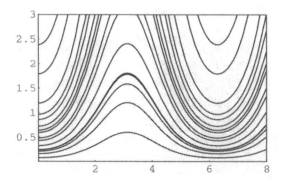

A very common mistake is to use = instead of == in a *Mathematica* equation. This can have disastrous effects, since if one uses, say, x′[t] = t in a call to a command that solves an equation, *Mathematica* will define x′[t] to be t; simply going back and adding the forgotten equal sign does not help, since x′[t] has been set to t and now x′[t] == t registers as True! So we have incorporated a warning.

```
VisualDSolve[x'[t] = x[t], {t, 0, 1}, {x, -3, 3}];
VisualDSolve::badinput: x'[t] = x[t] has = where it should have ==.
```

1.3 ▮ Options

There are many options to the commands in the VisualDSolve package. You can see all the default settings for a particular command by executing Options[*command*].

```
Options[VisualDSolve]

                       1
{AspectRatio -> ────────────, Axes -> True,
                 GoldenRatio
  Background -> GrayLevel[1], ComputeWindow -> Automatic,
  ContourLines -> False, ContourShading -> False,
  DefaultColor -> Automatic, DirectionField -> False,
  FastPlotting -> False, FieldMeshSize -> 15, InflectionCurve -> False
  InflectionStyle -> {AbsoluteDashing[{5, 5}]},
  InitialPointStyle -> {GrayLevel[0], PointSize[0.021]},
  InitialValues -> {}, Isoclines -> False, IsoclinePlotPoints -> 20,
  IsoclineShading -> False, IsoclineStyle -> {}, IsoclineValues -> {0
  MaxBend -> 10, MaxSteps -> 500, Method -> Automatic,
  PlotPoints -> 25, PlotStyle -> {}, PrecisionGoal -> Automatic,
  Rainbow -> False, SaveLastPoint -> False, SaveSolution -> False,
  ShowInitialValues -> False, SolutionName -> Solution,
  StayInWindow -> False, SymbolicSolution -> False,
  WindowShade -> None, WorkingPrecision -> 16}
```

Note that some of these are options specific to VisualDSolve, while others are built-in options to Plot or NDSolve, which can be passed through VisualDSolve; examples follow. To learn a single default setting, proceed as follows.

```
Options[VisualDSolve, InitialPointStyle]

{InitialPointStyle -> {GrayLevel[0], PointSize[0.021]}}
```

And for information about what an option accomplishes, ask for its usage statement.

```
?InitialPointStyle
```

InitialPointStyle is an option to VisualDSolve and related
    functions that specifies the style of the initial-value
    points. It can be a list or a single graphics primitive.

```
?StayInWindow
```

StayInWindow is an option to PhasePlot and VisualDSolve that,
    when True, causes the plots to stop when they hit the window
    frame (or the max value in the t-iterator, or the setting of
    ComputeWindow). It also causes the orbits to expand in both
    directions from the initial point(s). When this setting is
    used, Method must be set to the default of Automatic. When
    StayInWindow is False, then the specified t-values are used.
    If any initial values come with specific t-limits, then they
    override a StayInWindow request. This allows specific t-limits
    to be used for periodic solutions that will never leave the
    window. The exit data, in a form suitable for later use with
    InitialValues, is returned as output.

All usage messages for objects in the VisualDSolve package can be found in the Appendix.

1.4 ■ Setting and Seeing the Initial Values

The ShowInitialValues option allows us to see the initial values.

```
VisualDSolve[x'[t] == x[t], {t, -1, 1}, {x, -2, 3},
    InitialValues -> Grid[4], ShowInitialValues -> True];
```

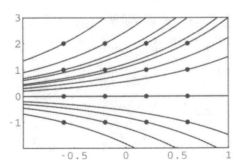

While a grid can be useful for a first look, usually it will be more convenient to specify the initial values exactly. We can specify one or more individual initial values, and we can change the style of the points placed at the initial values.

```
VisualDSolve[x'[t] == x[t], {t, -2, 2}, {x, 0, 8},
   ShowInitialValues -> True, InitialValues -> {0, 1},
   InitialPointStyle -> {GrayLevel[0.6], AbsolutePointSize[10]}];
```

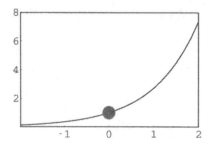

```
VisualDSolve[x'[t] == x[t], {t, -2, 2}, {x, 0, 8},
   ShowInitialValues -> True, InitialValues -> {{0, 1}, {0, 2}}];
```

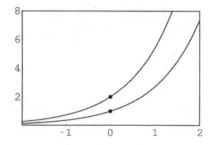

The next example shows how to specify a line of initial values; note how two of them are outside the window, but the visible part of the solutions show up properly.

```
VisualDSolve[x'[t] == x[t], {t, -1, 2}, {x, 0, 8},
   ShowInitialValues -> True,
   InitialValues -> Table[{0, i}, {i, 1, 10}]];
```

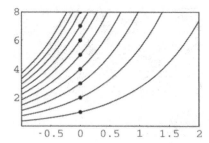

In fact, several of the solutions go way outside the window; we can see them by redrawing the image with a larger plot range.

```
Show[%, PlotRange -> {0, 16}];
```

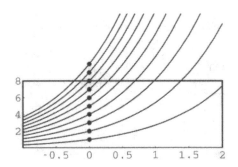

The StayInWindow option automatically cuts off computations if they ever leave the window. This clearly avoids a lot of unnecessary computation. However, it also slows down part of the computation, because the differential equation solver must constantly check to see if the window has been exited. Thus some experimentation might be needed in a specific case to learn which is faster.

When StayInWindow is used, then the output includes data about the window exit times. Each data point consists of an initial value followed by the t-bounds for the solution in the window. If the same equation is to be solved again, this data can be used as a setting to InitialValues, which will speed things up since StayInWindow can be turned off (an example of this is given at the end of this section). Since we want to cut off the computation outside the window, we restrict the initial values to the window.

```
VisualDSolve[x'[t] == x[t], {t, -1, 2}, {x, 0, 8},
    ShowInitialValues -> True, StayInWindow -> True,
    InitialValues -> Table[{0, i}, {i, 1, 7}]];
```

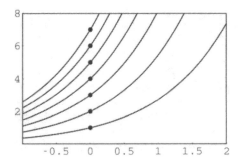

```
{-Graphics-, {{0, 1., {-1., 2.}},
    {0, 2., {-1., 1.38629}}, {0, 3., {-1., 0.980816}},
    {0, 4., {-1., 0.692569}},
    {0, 5., {-1., 0.470002}},
    {0, 6., {-1., 0.287502}}, {0, 7., {-1., 0.133531}}}}}
```

The following image confirms that the computations have shut off properly.

```
Show[%[[1]], PlotRange -> All];
```

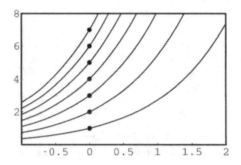

One might want the computation window to be different than the viewing window. For example, one might wish to not terminate for brief exits of the viewing window since such a solution might return to the window. The ComputeWindow setting allows for this. It can be used only when StayInWindow is turned on. In the next example the computation persists beyond the first window exit, but stops at the second one.

```
VisualDSolve[x'[t] == Cos[t] - t Sin[t],
    {t, 0, 7 Pi}, {x, -16, 7},
    InitialValues -> {0, 0}, StayInWindow -> True,
    ComputeWindow -> {-16, 15}];
```

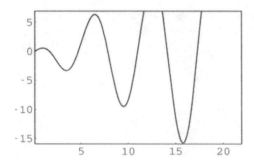

The InitialValues option allows the user to specify different *t*-intervals for different initial values. If no special *t*-interval is specified then the interval in the main iterator is used. This extra control can be useful in avoiding unnecessary computation in cases where a solution is known to go way outside the window.

```
VisualDSolve[x'[t] == Cos[t],
   {t, -0.5, 2 Pi}, {x, -2, 2},
   ShowInitialValues -> True,
   InitialPointStyle -> AbsolutePointSize[5],
   InitialValues -> {{0, -0.5},
                     {0, 0, {0, Pi}},
                     {1, 1.5, {2, 5}} }];
```

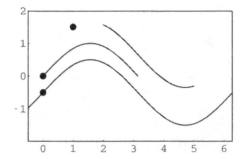

To illustrate the speedup when exact *t*-domains are used, we first generate eight solutions to a differential equation with `StayInWindow` turned on.

```
VisualDSolve[x'[t] == x[t], {t, -1, 2}, {x, 0, 8},
   InitialValues -> Table[{0, i}, {i, 0, 7}],
   StayInWindow -> True,
   DisplayFunction -> Identity] //Timing
```

```
{1.9 Second, {-Graphics-,
    {{0, 0, {-1., 2.}}, {0, 1., {-1., 2.}},
     {0, 2., {-1., 1.38629}}, {0, 3., {-1., 0.980816}},
     {0, 4., {-1., 0.692569}},
     {0, 5., {-1., 0.470002}},
     {0, 6., {-1., 0.287502}}, {0, 7., {-1., 0.133531}}}
    }}
```

When we redo the computation using the computed *t*-limits there is a substantial speedup (over 30%; all timings are for a PowerMac 8100).

```
VisualDSolve[x'[t] == x[t], {t, -1, 2}, {x, 0, 8},
   InitialValues -> %[[2, 2]],
   DisplayFunction -> Identity]; //Timing //First
```

```
1.31667 Second
```

1.5 ▪ Style and Accuracy Control

We can specify distinct styles for solution curves with the `PlotStyle` option. But note that each style list must be wrapped in its own sublist; forgetting this is a common mistake. Using, say, `PlotStyle -> {Blue, Red, Green}` to get three distinct colors for three curves won't work. Neither will `PlotStyle -> {Blue, Thickness[0.02],` `Dashing[{0.02}]}`; what happens here is that the list of attributes will apply to all curves. To repeat: Each style list should be in its own sublist, and there should be a sublist for each curve.

```
VisualDSolve[x'[t] == x[t], {t, -1, 2}, {x, 0, 8},
    ShowInitialValues -> True, InitialValues -> Table[{0, i}, {i, 1, 3}],
    PlotStyle -> {{Dashing[{0.02, 0.04}], Thickness[0.012]},
                  {Thickness[0.012], GrayLevel[0.5]}, {}}];
```

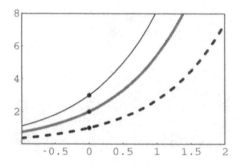

The attributes such as `Thickness` and `Dashing` interpret their arguments as percentages of the entire plot. If you wish to specify exact thicknesses in terms of points (a point is 1/72 of an inch), use `AbsoluteThickness` or `AbsoluteDashing`. Because the `VisualDSolve` package loads the `Graphics`Colors`` package, we can use words such as `Black`, `Gray` (same as `GrayLevel[0.75]`), `Red`, or `LemonChiffon` for colors. See section 1.1 (page 4 for information on the full list of colors).

```
VisualDSolve[x'[t] == x[t], {t, -1, 2}, {x, 0, 8},
    ShowInitialValues -> True, InitialValues -> Table[{0, i}, {i, 1, 3}],
    PlotStyle -> {{Gray, AbsoluteThickness[3]},
                  {GrayLevel[0.4], AbsoluteThickness[2]},
                  {AbsoluteThickness[1]}}, AxesLabel -> {t, x}];
```

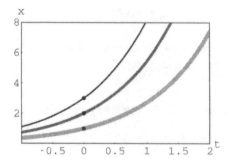

If one wants all the curves to be in a certain style then one can feed a single attribute or a single list of attributes to PlotStyle. We can also use PlotLabel to pass along a label for the plot.

```
VisualDSolve[x'[t] == x[t], {t, -1, 2}, {x, 0, 8},
  InitialValues -> Table[{0, i}, {i, 1, 3}],
  PlotStyle -> {Gray, AbsoluteThickness[3]},
  PlotLabel -> "Three solutions to dx/dt = x[t]"];
```

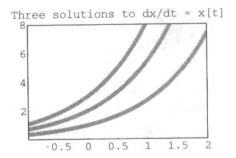

Rainbow is a separate option that causes different hues to be used for the different solution curves. Any styles (even colors) specified by PlotStyle are still respected. The WindowShade option causes a shaded rectangle to appear behind the window, which enhances plots that use color; its setting must be a color or gray level. The output of the following command is in color plate 1.

```
VisualDSolve[x'[t] == x[t], {t, -1, 2}, {x, 0, 8},
  InitialValues -> Table[{0, i}, {i, 1, 5, 0.5}],
  ShowInitialValues -> True, WindowShade -> Gray,
  InitialPointStyle -> {AbsolutePointSize[4], White},
  PlotStyle -> AbsoluteThickness[2], Rainbow -> True];
```

The default setting of the MaxSteps option to NDSolve is a mere 500 (1000 in version 3.0). This is not sufficient for many examples, and you may have to step it up quite a bit. Of course, you must be aware of the memory limitations on your machine.

```
VisualDSolve[x'[t] + t Cos[Pi t]^2 x[t] == t Cos[t],
  {t, 0, 50}, {x, -11, 11}, InitialValues -> {0, 0}];
```

NDSolve::mxst: Maximum number of 500 steps reached at the point 12.0407.

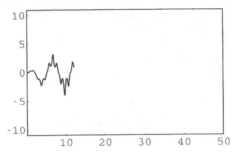

```
VisualDSolve[x'[t] + t Cos[Pi t]^2 x[t] == t Cos[t],
  {t, 0, 50}, {x, -11, 11},
  InitialValues -> {0, 0}, MaxSteps -> 3500];
```

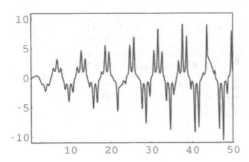

You can also pass along other NDSolve options, such as AccuracyGoal, WorkingPrecision, and PrecisionGoal (see sections 2.3 and 15.5 for examples).

Complicated solution curves such as this one illustrate an important point. The default plotting method in *Mathematica* uses 25 plot points to get started and tries to make the angles between successive segments less than 10°. But one can pass whatever Plot options one likes through VisualDSolve. By tweaking these settings, we get a finer picture of this complicated function.

```
VisualDSolve[x'[t] + t Cos[Pi t]^2 x[t] == t Cos[t],
  {t, 0, 50}, {x, -11, 11},
  InitialValues -> {0, 0}, MaxSteps -> 3500,
  PlotPoints -> 100, MaxBend -> 10];
```

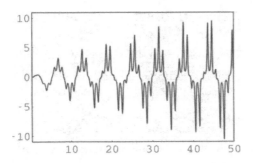

We have implemented an option called FastPlotting that speeds up plotting, cuts down on memory requirements, and often produces more accurate curves! To understand the principle underlying FastPlotting one must have a rudimentary understanding of how *Mathematica*'s numerical differential equation solver, NDSolve, works. NDSolve computes many (hundreds, or even thousands) points on the solution curve and then turns those points into an interpolating function. Since the points are available from the InterpolatingFunction object, one could try to get a curve by simply connecting these points with straight lines. This is not what Plot does; Plot invokes an adaptive algorithm, and so requires evaluating the interpolating function at, perhaps, one hundred or two hundred points. But it turns out that, often, the points inherent in the definition of the interpolating function are already very good. After all, some adaptive work takes place because NDSolve will tend to work hard where the solution curves sharply, thus causing more points to be generated in the vicinity of a sharp bend.

On the other hand, the interpolating function might involve 10,000 points, which is overkill as far as plotting is concerned. Attempting to connect so many points would create an object that took up a tremendous amount of memory. In such a case the default method of using Plot is the right way to proceed, probably with an increased value of PlotPoints. If FastPlotting is used then PlotPoints has no effect. Finally, FastPlotting cannot be used if SymbolicSolution is set to True. Here is an example. First we use the default method, with 36 initial values.

```
VisualDSolve[x'[t] == 3 x[t] Sin[x[t]] - t,
  {t, 0, 10}, {x, -5, 5},
  InitialValues -> Grid[6]]; //Timing //First
```

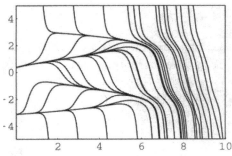

13.7667 Second

Now we turn to FastPlotting.

```
VisualDSolve[x'[t] == 3 x[t] Sin[x[t]] - t,
  {t, 0, 10}, {x, -5, 5},
  FastPlotting -> True, InitialValues -> Grid[6]]; //Timing //First
```

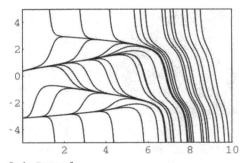

9.1 Second

FastPlotting is faster, takes less memory, and, as is evident upon close inspection, provides a better image. To get such a good image using the default plotting method, we have to increase the PlotPoints to 100, which increases both time and memory requirements.

```
VisualDSolve[x'[t] == 3 x[t] Sin[x[t]] - t,
    {t, 0, 10}, {x, -5, 5},
    PlotPoints -> 100, InitialValues -> Grid[6]]; //Timing //First
```

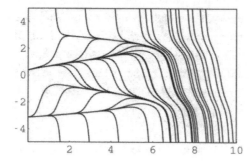

17.2667 Second

Note that FastPlotting is inappropriate in a situation where NDSolve needs very few points. For example, a request to solve $dx/dt = 2t$ puts very little strain on NDSolve; it generates only 14 points, and a too-straight curve results if we just connect the interpolating points.

```
VisualDSolve[x'[t] == 2 t,
    {t, 0, 3}, {x, -1, 9},
    InitialValues -> {0, 0}, FastPlotting -> True];
```

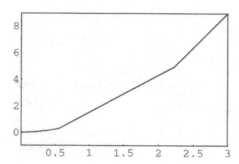

If the user wishes to use FastPlotting as a default, that can be arranged for the appropriate functions in the package as follows.

```
SetOptions[{VisualDSolve, PhasePlot, SystemSolutionPlot},
    FastPlotting -> True];
```

1.6 ▮ **Symbolic Solutions**

We can ask `VisualDSolve` to make an attempt at finding a symbolic solution. If it succeeds, then it will use the symbolic solution to plot the curves and output the general solution as a rule, using C for the constant.

```
VisualDSolve[x'[t] == Cos[t] - x[t],
    {t, 0, 2 Pi}, {x, -1, 1},
    SymbolicSolution -> True,
    InitialValues -> {{0, 0}, {0, 0.5}}]
```

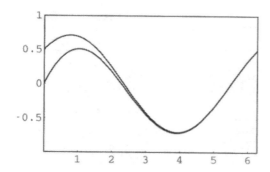

$$\{-\text{Graphics}-, \ \{\{x[t] \ \to \ \frac{C}{E^t} + \frac{\text{Cos}[t] + \text{Sin}[t]}{2}\}\}\}$$

If you wish to use such a solution, you can get at it as follows.

```
x[t] /. First[Last[%]]
```

$$\frac{C}{E^t} + \frac{\text{Cos}[t] + \text{Sin}[t]}{2}$$

As a reminder, note that if you want to solve a differential equation symbolically but do not wish to plot the solution, then you can use the built-in `DSolve` as follows. More information about using `DSolve` and `NDSolve` is given in Chapter 6.

```
DSolve[x'[t] == Sin[t] x[t], x[t], t]
```

$$\{\{x[t] \ \to \ \frac{C[1]}{E^{\text{Cos}[t]}}\}\}$$

Of course, many problems do not admit symbolic solutions. If the symbolic attempt fails, then a message is returned and the plot is generated numerically.

```
VisualDSolve[x'[t] == Sqrt[Sin[t]],
  {t, 0, 3}, {x, 0, 4},
  SymbolicSolution -> True,
  InitialValues -> {{0, 0}, {0, 0.5 }}];
```

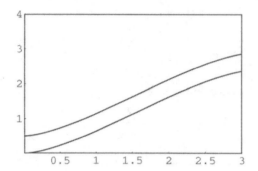

Sometimes the symbolic solution has several parts. When that happens, the symbolic solution is returned, but a numerical approach is used to get the plotted curve. Here is an example.

```
VisualDSolve[x'[t] == -t/x[t],
  {t, -3, 3}, {x, -3, 3},
  InitialValues -> {2, 2}, SymbolicSolution -> True]
```

VisualDSolve::manycs:
    The symbolic solution has more than one case, so
    numerical techniques are being used for the plotting.

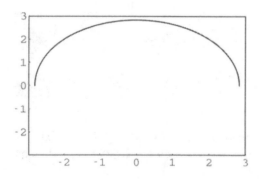

$\{-Graphics-, \{\{x[t] \rightarrow -Sqrt[C - t^2]\}, \{x[t] \rightarrow Sqrt[C - t^2]\}\}\}$

Of course, more information can be gleaned from a symbolic solution than a numerical one. Whether that information is important depends on the context. A complicated symbolic solution is often no more useful than a numerical approximation. But if one has a symbolic solution in hand and it disagrees with the results of the numerical method, that is a very important piece of information. Here's an example where that happens (due to Courtney Coleman, Harvey Mudd College).

```
VisualDSolve[x'[t] == 2 x[t] + Cos[t],
   {t, 0, 3 Pi}, {x, -1, 200},
   InitialValues -> {0, -2/5}];
```

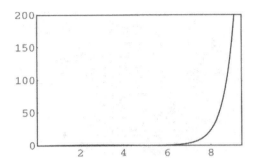

```
VisualDSolve[x'[t] == 2 x[t] + Cos[t],
   {t, 0, 3 Pi}, {x, -1, 1},
   InitialValues -> {0, -2/5}, SymbolicSolution -> True]
```

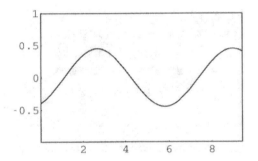

$$\{-\text{Graphics}-, \{\{x[t] \to C\ E^{2\ t} + \frac{-2\ \text{Cos}[t] + \text{Sin}[t]}{5}\}\}\}$$

Of course, it is an easy matter to check that the symbolic solution is correct (methods of verifying a solution, symbolic or numerical, are discussed in section 2.3).

```
soln = x[t] /. %[[-1, 1]]
```

$$C\ E^{2\ t} + \frac{-2\ \text{Cos}[t] + \text{Sin}[t]}{5}$$

```
D[soln, t] == 2 soln + Cos[t] //Simplify
True
```

The problem with the numerical approach in this case is the tremendous divergence that takes place. The reader can see this divergence by looking at the symbolic solution with slightly perturbed initial values.

1.7 ■ Direction Fields

The `DirectionField` option produces a background of slope lines for the differential equation.

```
VisualDSolve[x'[t] == 3 x[t] Sin[x[t]] - t,
   {t, 0, 10}, {x, -5, 5},
   DirectionField -> True, InitialValues -> Grid[4]];
```

Note that the lines are not placed on an exact rectangular grid. Rather, there are two grids, slightly offset; this produces a better visual effect. In Chapter 4 on phase plots, where the lengths of the lines vary, we will see the virtue of random placement of the lines, and using shapes other than lines (curvy fish, to be precise) to represent directions.

We can control the number of field lines; the default is 225, in roughly a 15×15 grid. The following example shows a 20×20 grid; using $\{m, n\}$ produces an $m \times n$ grid. We can also control the color of the field lines. In the example that follows we include a dark gray `WindowShade` and ask for white field lines.

```
VisualDSolve[x'[t] == 3 x[t] Sin[x[t]] - t,
   {t, 0, 10}, {x, -5, 5},
   DirectionField -> True, FieldMeshSize -> 20,
   FieldColor -> White, WindowShade -> GrayLevel[0.3],
   InitialValues -> Grid[4]];
```

The default setting of the length of the field lines is 0.8, which allows enough space so that lines do not crash. This can be changed.

```
VisualDSolve[x'[t] == 3 x[t] Sin[x[t]] - t,
    {t, 0, 10}, {x, -5, 5}, DirectionField -> True,
    FieldMeshSize -> 10, FieldLength -> 1.3];
```

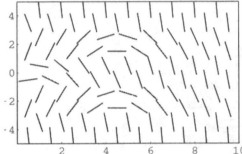

If we know that we want the image to have a certain aspect ratio, then we can include that setting, and the resulting image will have lines that appear to be of the same length. However, if we change the aspect ratio (by deselecting the **PreserveAspectRatio** menu choice and manually changing the graphic's frame), then the line-lengths in the image can change.

```
VisualDSolve[x'[t] == x[t], {t, 0, 10}, {x, -5, 5},
    DirectionField -> True, FieldMeshSize -> 10,
    AspectRatio -> 1/5];
```

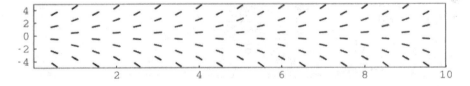

1.8 ■ Isoclines

Direction fields provide a way of gaining information about a differential equation without actually looking at solutions. Another way of getting such information is via isocline curves. An isocline curve for a value c is the curve connecting all points where the slope of the solution curve through that point is c. Typically c is 0, in which case the isocline is called a *nullcline*; we have set the default of IsoclineValues to be just {0}. If you want to see only the isocline, then that can be done by a simple ContourPlot. Consider the example $x' = 3x\sin(x) - t$; it is straightforward to use ContourPlot to see the nullcline.

```
ContourPlot[3 x Sin[x] - t, {t, 0, 10}, {x, -5, 5},
    Contours -> {0}, ContourShading -> False, PlotPoints -> 50];
```

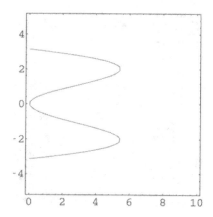

VisualDSolve has the ability to place the isoclines as a background to the solution curves. In the example below we specify a thick style for the nullcline, to distinguish it from the solution curves; and we increase the PlotPoints setting to get a smoother nullcline (to avoid ambiguity, this type of plot point setting is called IsoclinePlotPoints).

```
VisualDSolve[x'[t] == 3 x[t] Sin[x[t]] - t,
    {t, 0, 10},  {x, -5, 5},
    Isoclines -> True,
    IsoclineStyle -> AbsoluteThickness[3],
    IsoclinePlotPoints -> 50, InitialValues -> Grid[4],
    AxesLabel -> {t, x}];
```

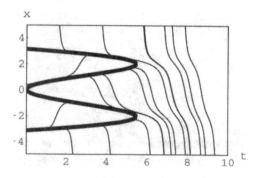

You can see that each solution curve has a horizontal tangent wherever it touches the nullcline. Thus the nullcline picks out all the critical points on all the solutions. Here is an example in gray shades that shows some flat crossings more clearly. Of course, such diagrams can often be made very clear by a judicious choice of bright colors.

```
VisualDSolve[x'[t] == 3 x[t] Sin[x[t]] - t,
   {t, 0, 8},  {x, -5, 5},
   InitialValues -> Table[{x, -5}, {x, 5.5, 8, 0.5}],
   Isoclines -> True, IsoclinePlotPoints -> 50,
   IsoclineStyle -> {White, AbsoluteThickness[2]},
   PlotStyle -> AbsoluteThickness[2], FastPlotting -> True,
   AxesLabel -> {t, x}, WindowShade -> GrayLevel[0.5]];
```

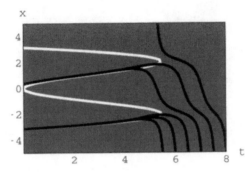

The nullclines give us a very important piece of information. To understand it more fully, look at what happens when we show the nullcline together with a direction field.

```
VisualDSolve[x'[t] == 3 x[t] Sin[x] - t,
   {t, 0, 8},  {x, -5, 5},
   FieldMeshSize -> 25, DirectionField -> True,
   Isoclines -> True, IsoclinePlotPoints -> 50];
```

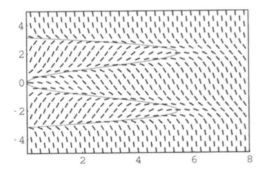

As you can see, the nullcline separates rising slopes from falling slopes. In other words, on one side of the nullcline, solutions decrease, on the other side they increase. We can use this to help our understanding of the solution curves by suppressing the visually cluttered direction field and simply shading the regions differently. This is done via the `IsoclineShading` option. When using this it seems redundant to have the actual isocline curve shown; we suppress it by setting `ContourLines -> False`.

```
VisualDSolve[x'[t] == 3 x[t] Sin[x[t]] - t,
   {t, 0, 10}, {x, -5, 5},
   Isoclines -> True, IsoclineShading -> True,
   IsoclineStyle -> {GrayLevel[0.5], AbsoluteThickness[3]},
   Contours -> 10, PlotStyle -> AbsoluteThickness[2],
   InitialValues -> Grid[4], ContourLines -> False,
   IsoclinePlotPoints -> 50];
```

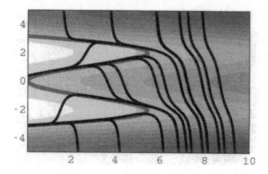

One can ask for more isocline values by including them specifically in the setting of IsoclineValues; or one can set IsoclineValues to be a single integer, such as 10, in which case 10 equally spaced isocline values will be used. This sort of image has been used by some experts in differential equations (*e.g.*, [HW, p. 16]).

```
VisualDSolve[x'[t] == 3 x[t] Sin[x[t]] - t,
   {t, 0, 8},  {x, -5, 5},
   InitialValues -> Table[{i, -5}, {i, 5.5, 8, 0.5}],
   Isoclines -> True, IsoclineStyle -> White,
   IsoclineValues -> 10, IsoclinePlotPoints -> 50,
   WindowShade -> GrayLevel[0.5], AxesLabel -> {t, x}];
```

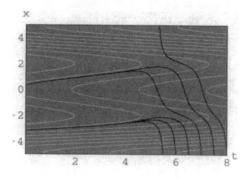

This sort of view is better if we shade the isocline regions, as in the next example. Note that if IsoclineShading is turned on, then there is no need to use the Isoclines option. Also we use a setting of 15 for IsoclineValues; we could ·also have used Contours -> 15, since that will be passed through to the relevant ContourPlot command. The AspectRatio -> Automatic setting causes the axes scales to match the actual axis-lengths (an aspect ratio of 1 in this case).

```
VisualDSolve[x'[t] == - t x[t], {t, -5, 5}, {x, -5, 5},
   InitialValues -> Table[{-i, i}, {i, -5, 5, 0.5}],
   IsoclineShading -> True, IsoclineValues -> 15,
   IsoclinePlotPoints -> 50,
   AxesLabel -> {t, x}, AspectRatio -> Automatic];
```

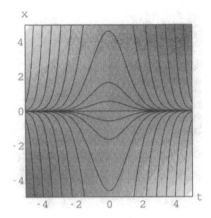

Suppose one wants such a picture, but with a special style for the nullcline. We can do that by using `Contours` to specify the contour values for the shaded part, but `IsoclineValues` to specify the actual values for a curves-only isocline plot. In other words, we have arranged that the shaded plot first look to any options for `ContourPlot` that are passed, while an unshaded isocline plot—indicated by `Isoclines -> True`—uses the settings in `IsoclineValues`. This seems complicated, but it allows us to get an image as follows, which combines a single nullcline plot with a background that is the shaded isocline plot. Because the default setting of `IsoclineValues` is in fact `{0}`, which is just what we want, we can omit it.

```
VisualDSolve[x'[t] == 3 x[t] Sin[x[t]] - t,
   {t, 0, 10}, {x, -5, 5}, Isoclines -> True,
   IsoclineStyle -> {GrayLevel[0.5], AbsoluteThickness[3]},
   IsoclineShading -> True, Contours -> 10,
   PlotStyle -> AbsoluteThickness[2], InitialValues -> Grid[4],
   ContourLines -> True, IsoclinePlotPoints -> 50];
```

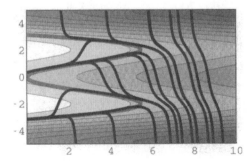

Here is another example, one that is pretty simple to understand because the nullcline is just the unit circle; thus the solutions descend in the unit disk and ascend outside it. Note that one can use the variable *t* in the `Table` that computes the initial values. The color output of the following command is in color plate 2.

```
VisualDSolve[x'[t] == x[t]^2 + t^2 - 1,
  {t, -2.5, 2.5}, {x, -2.5, 2.5},
  InitialValues ->
    Table[1.5 {Cos[t], Sin[t]}, {t, 0, 2 Pi, Pi/10}],
  ShowInitialValues -> True, StayInWindow -> True,
  IsoclineShading -> True,
  IsoclineStyle -> AbsoluteThickness[2],
  Rainbow -> True, PlotStyle -> AbsoluteThickness[1.5],
  Contours -> 15, Isoclines -> True,
  IsoclinePlotPoints -> 50,
  AspectRatio -> Automatic];
```

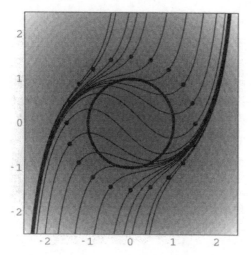

Here is a complicated example taken from [HW, p. 40]. The light colors correspond to maximum steepness, the dark colors to minimum steepness (*i.e.*, maximum negative steepness). The high setting for `IsoclinePlotPoints` makes this a memory-intensive example. As before, we add the nullclines in black; note how they connect the critical points on the solutions. The output of the following is in color plate 3.

```
VisualDSolve[x'[t] == Sin[t * x[t]], {t, -1, 6}, {x, -1, 7},
   InitialValues -> Table[{0, x}, {x, -0.75, 6, 0.5}],
   IsoclineShading -> True, Contours -> 10, IsoclinePlotPoints -> 100,
   FastPlotting -> True, Isoclines -> True, Rainbow -> True,
   PlotStyle -> AbsoluteThickness[1.5]];
```

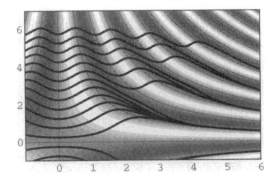

Here is another example that illustrates the shading and coloring features of `VisualDSolve`. The equation is an example of a Riccati equation. Again, note how the nullcline (thick white line) separates the region of positive slope from the region of negative slope. Note also how the upper half of the nullcline (parabola) repels solutions while the lower half attracts them.

```
VisualDSolve[x'[t] == x[t]^2 - t, {t, -5, 20}, {x, -10, 10},
   ShowInitialValues -> True, StayInWindow -> True,
   Isoclines -> True, IsoclineStyle -> {AbsoluteThickness[5], White},
   InitialValues -> { {6.4, 0.8}, {-4.3, -10}, {-2.7, -10},
      {-2.1, -10},  {-1.7, -10},  {-1.3, -10}, {-0.35, -10},
      {1.1, -10},   {4.5, -10},   {6.8, -10},  {10, -10},
      {12.4, -10},  {15, -10},    {19.1, 10},  {15.9, 10},
      {12.6, 10},   {8.8, 10},    {5.6, 10},   {3., 10},
      {16.6, 0.38}, {13.7, 0.38}, {10.2, 0.8}, {6.4, 0.8}},
   PlotStyle -> AbsoluteThickness[1], AxesLabel -> {t, x},
   IsoclineShading -> True, Contours -> 10,
   IsoclinePlotPoints -> 50,
   PlotLabel -> "Solution curves for x' = x^2 - t"];
```

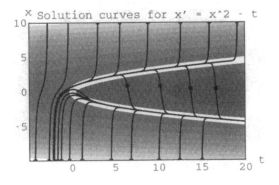

1.9 ■ Inflection Curves

One can also look at the inflection curves, which show the places where the solutions have inflection points. Here's an example from [BCB, p. 7]. The default is to dash the inflection curve; naturally, one can control this with the `InflectionStyle` option.

```
VisualDSolve[x'[t] == -Sin[t] x[t] + Cos[t],
  {t, -10, 10}, {x, -10, 10},
  InitialValues -> Table[{-10, i}, {i, -10, 10, 2}],
  InflectionCurve -> True, IsoclinePlotPoints -> 100];
```

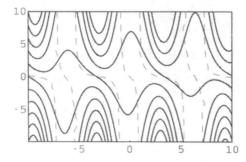

1.10 ■ Controlling the Numerical Method

The built-in numerical solver, `NDSolve`, uses a sophisticated adaptive numerical algorithm. For pedagogical purposes, we have allowed the user to specify that either Euler's method or the classical fourth-order Runge–Kutta method be used. For a simple example, consider $x' = x + \sin t$. Note that the solution image is not simply piecewise linear based on the Euler data, but is an interpolating function based on that data. For this example, we have already used *Mathematica* to find the exact solution for us in section 1.6 (it is $x(t) = e^{-\cos t}$); thus we can use the `Epilog` option to add a point at the true solution value when $t = 8\pi$.

```
t1 = 8 Pi;
VisualDSolve[x'[t] == Sin[t] * x[t],
  {t, 0, 8 Pi}, {x, -0.5, 6},
  InitialValues -> {0, 1}, Method -> Euler, EulerSteps -> 50,
  Epilog -> {AbsolutePointSize[5], Point[{t1, E^(-Cos[8 Pi])}]}];
```

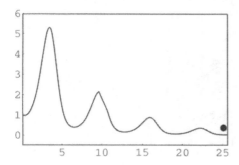

Of course, it is nice in these situations to superimpose a plot of the true solution.

```
Plot[E^(-Cos[t]), {t, 0, t1},
   DisplayFunction -> Identity,
   PlotStyle -> {{AbsoluteThickness[2], GrayLevel[0.5]}}];

Show[%%, %, DisplayFunction -> $DisplayFunction];
```

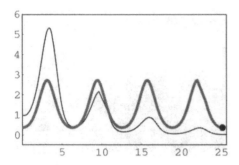

One could make an animation showing how the increase in the number of steps leads to convergence. We will do that in a moment for the Runge–Kutta method. But as an aside, let us find out how many steps *Mathematica* requires to get its numerical solution (using its default precision goal of 6, which means that it strives at each step for 6 significant digits; more precisely, it tries to keep the local (relative) error to less than 10^{-6}). We use a couple of tricks to find this out. We first set a counter to 0; then we define f so that it increments the counter each time it is called. But this is not sufficient, because NDSolve would, in such a case, look at f only once and then rewrite its expression internally. We can override this by specifying that the definition of f applies only to numbers (via attaching _Real to the dummy variables of f). This does the job and we see that counter moves up to 600, indicating that the equation was looked at 600 times.

```
counter = 0;
f[x_Real, t_Real] := (counter++; x Sin[t])

VisualDSolve[x'[t] == f[x[t], t], {t, 0, 8 Pi}, {x, -0.5, 8},
   InitialValues -> {0, 1}, DisplayFunction -> Identity];

counter
```

600

If we use Euler's method with 600 steps, we get some improvement over the 50-step method, but the error at the end is still quite visible.

```
t1 = 8 Pi;
VisualDSolve[x'[t] == Sin[t] * x[t], {t, 0, t1}, {x, -0.1, 7},
   InitialValues -> {0, 1}, Method -> Euler, EulerSteps -> 600,
   Epilog -> {AbsolutePointSize[5], Point[{t1, E^(-Cos[t1])}]}];
```

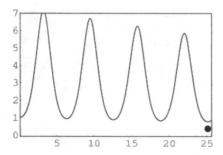

To illustrate the Runge–Kutta method we use a more complicated example [BCB, p. 73]. First, let us look at the default solution. We tweak the PlotPoints option (as well as MaxSteps) for finer resolution, as explained on page 16; and we give the name trueSoln to this image.

```
trueSoln = VisualDSolve[x'[t] + t Cos[Pi t]^2 x[t] == t Cos[t],
   {t, 0, 20}, {x, -10, 10}, InitialValues -> {0, 0},
   MaxSteps -> 1000, FastPlotting -> True];
```

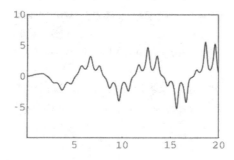

```
VisualDSolve[x'[t] + t Cos[Pi t]^2 x[t] == t Cos[t],
   {t, 0, 20}, {x, -10, 10},
   Method -> RungeKutta4, RKSteps -> 60,
   InitialValues -> {0, 0},  PlotPoints -> 100];
```

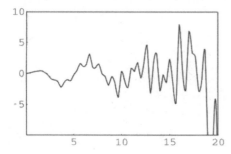

In order to get a feel for the effect of increasing the number of Runge–Kutta steps, we can make an animation. We can define RK[*n*] to create the image corresponding to *n* steps.

```
RK[n_] := VisualDSolve[x'[t] + t Cos[Pi t]^2 x[t] == t Cos[t],
          {t, 0, 20}, {x, -10, 10},
          InitialValues -> {0, 0}, Method -> RungeKutta4,
          RKSteps -> n, DisplayFunction -> Identity];
```

We can now show simultaneously the true solution and the Runge–Kutta approximation. We can also use the PlotLabel option (this is an option to Plot, but VisualDSolve just passes it along) to create a label that shows clearly the number of Runge–Kutta steps in a specific image; this sort of detail makes an animation much more informative. While PlotLabel -> n would do the job, StringForm allows us to build up a string with embedded values at the places where " appears. Here's an example:

```
a = apples; b = oranges;
StringForm["You can't compare " and "!", a, b]
```

You can't compare apples and oranges!

```
n = 30;
Show[RK[n], DisplayFunction -> $DisplayFunction,
    PlotLabel -> StringForm["" steps", n]];
```

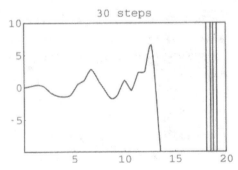

For more font control (convenient for lecture presentations) use FontForm to change the size of the plot label.

```
n = 30;
Show[RK[n],
  DisplayFunction -> $DisplayFunction,
  PlotLabel -> FontForm[StringForm["`` steps", n],
    {"Courier-Bold", 16}]];
```

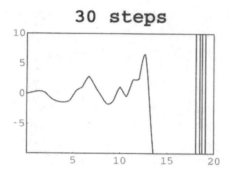

We can now easily use a Do loop to create an animation that shows the effect of increasing the number of steps. The code that follows generates a 14-frame animation.

```
Do[
  Show[trueSoln, RK[n], PlotLabel -> StringForm["`` steps", n],
    DisplayFunction -> $DisplayFunction],
  {n, 30, 210, 15}]
```

We cannot show an animation on the printed page, so instead we use a single image that contains several frames. This is done via the GraphicsArray command, which we now illustrate. We first compute and store the images in a variable called image. We then feed the image values as a matrix to GraphicsArray. Of course, there is ample opportunity for varying the colors and thicknesses to make this example easier to grasp and more visually appealing.

```
Do[image[n] = Show[trueSoln, RK[n],
    PlotLabel ->
      FontForm[StringForm["Runge-Kutta with `` steps", n],
        {"Times-Roman", 9}],
    DisplayFunction -> Identity], {n, 20, 120, 20}]
Show[GraphicsArray[{{image[20],  image[40]},
                    {image[60],  image[80]},
                    {image[100], image[120]}}],
  DisplayFunction -> $DisplayFunction];
```

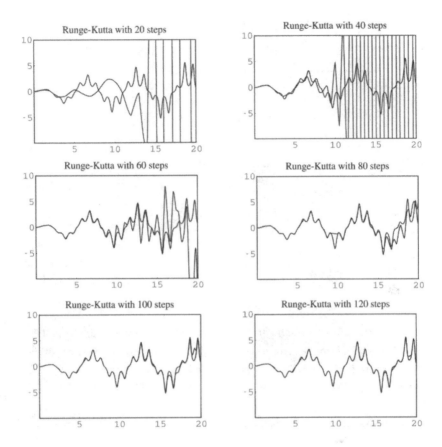

1.11 ■ Using the Output

VisualDSolve creates a plot, but then discards the information used in generating that plot. A user might wish to use the function or functions that represent the solution or solutions in other ways, without having to waste time regenerating them. We have provided this capability through the SaveSolution option. When this is set to True, the solutions are saved in a global variable called Solution. The user can then access this and use it as he or she chooses.

```
VisualDSolve[x'[t] + t Cos[Pi t]^2 x[t] == t Cos[t],
   {t, 0, 20}, {x, -10, 10}, InitialValues -> {0, 0},
   SaveSolution -> True, MaxSteps -> 2000];
```

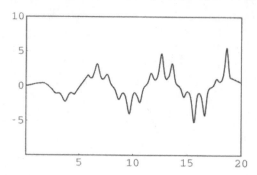

```
Short[Solution]
```

```
InterpolatingFunction[{0., 20.}, <>][t]
```

Warning: *Mathematica* can take an inordinately long time to display an `Inter-polatingFunction`. So either avoid such displays, or use `Short`, which provides the same output as when `Short` is not used, but much more quickly! Note that the solution is an expression in t. We can make a table of values.

```
Table[Solution, {t, 0, 10, 1}]
```

```
{0., 0.321559, 0.122717, -1.07115, -1.46933, -0.246941,
   1.25117, 1.44136, 0.222423, -1.12704, -1.32495}
```

We can even differentiate the solution and plot or tabulate that! So why not check to see how well it satisfies the differential equation?

```
xDer = D[Solution, t];
Table[N[xDer + t Cos[Pi t]^2 Solution - t Cos[t]],
   {t, 0, 10, 1}]
```

$$\{0.00117886, -7.18817 \ 10^{-6}, 5.11412 \ 10^{-7}, 1.02916 \ 10^{-6},$$
$$0.0000174989, -7.62465 \ 10^{-6}, 0.0000394303, -0.000066045,$$
$$9.67588 \ 10^{-6}, 0.0000312287, 1.52082 \ 10^{-6}\}$$

Not bad. We-see a maximum error of about 10^{-5}; this is consistent with the default goals of `NDSolve`, which is what `VisualDSolve` uses. We can look at the error over the entire *t*-domain. Taking absolute values and base-10 logarithms seems reasonable.

```
Plot[Log[10, Abs[xDer + t Cos[Pi t]^2 Solution - t Cos[t]]],
     {t, 0, 20}, PlotRange -> {-8, -2}];
```

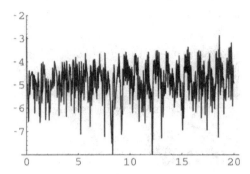

Note that an `InterpolatingFunction` contains some useful information. Since e in e[t] is called the head of e[t], we can get at it via Head[] (expr[[0]] is the same as Head[expr]). Then the first member of the interpolating function is the domain of the interpolating function. The second entry contains lots of data, but it is noteworthy that the length of this data set tells us how many interpolating points were used. For interpolating functions that come from differential equations, this tells us the number of steps that the numerical solver used.

```
intFunc = Head[Solution];
intFunc[[1]]     (* use intFunc[[1, 1]] in version 3.0 *)

{0., 20.}


Length[intFunc[[2]]]    (* use intFunc[[3, 1]] in version 3.0 *)

907
```

We can try this on the example from section 1.10, where we used a counter to learn that the differential equation was looked at 600 times. A more appropriate name for the solution storage variable can be used by sending it, as a string, via the `SolutionName` option. When this is used, `SaveSolution` is automatically turned on.

```
VisualDSolve[x'[t] == x[t] Sin[t],
   {t, 0, 8 Pi}, {x, -0.5, 8},
   InitialValues -> {0, 1}, DisplayFunction -> Identity,
   SolutionName -> "lengthExample"];
```

`GetPts` is a utility in the `VisualDSolve` package that gets the points that define an `InterpolatingFunction` (this is discussed in more detail in section 2.4).

```
Length[GetPts[lengthExample]]

267
```

Note that this is way less than the 600 we got earlier. The reason is that NDSolve uses a convergence method requiring several evaluations of the function defining the differential equation before deciding on a single interpolation point. We are seeing here that the interpolant is based on 267 *t*-values, but we know from our earlier work that the differential equation was evaluated 600 times. This point also applies to the Runge–Kutta method. If RKSteps is set to, say, 200, then one would say that 200 steps were used; but each step requires looking at the equation four times, so there are really 800 "function evaluations." Either method can be used for comparison, but one must be consistent.

If the number of initial points is greater than 1, then SaveSolution stores a list with all the solutions.

```
VisualDSolve[x'[t] == x[t] Sin[t],
   {t, 0, 8 Pi}, {x, -0.5, 8},
   DisplayFunction -> Identity,
   InitialValues -> Table[{0, x}, {x, 1, 2, 0.2}],
   SolutionName -> "sixInitialValues"];

Length[sixInitialValues]

6
```

We can see the different complexities of the six data sets by "mapping" the function that gives the number of interpolating points—we use the pure function Length[#[[0, 2]]]&—onto the solution set (see page 35 for comments on doing this in version 3.0).

```
Map[Length[#[[0, 2]]] &, sixInitialValues]

{267, 247, 260, 229, 211, 202}
```

Sometimes one wants to keep not the whole function, but just the very last point. Then use the SaveLastPoint option, which saves the last value(s) in a variable called LastPoint. Of course, this information is available from the interpolating functions that we can save in Solution, but that would use up more memory than is necessary for this small bit of data.

```
VisualDSolve[x'[t] == x[t] Sin[t],
   {t, 0, 8 Pi}, {x, -0.5, 8},
   DisplayFunction -> Identity, SaveLastPoint -> True,
   InitialValues -> Table[{0, x}, {x, 1, 2, 0.2}]];

LastPoint

{0.99998, 1.19999, 1.39997, 1.59999, 1.8, 2.}
```

Suppose we wish to compare the results of the built-in solver with the fourth-order Runge–Kutta method. Focusing on the same example as in section 1.10, we first use the built-in method, which we already know used 600 evaluations.

```
VisualDSolve[x'[t] + t Cos[Pi t]^2 x[t] == t Cos[t],
  {t, 0, 20}, {x, -10, 10}, InitialValues -> {0, 0},
  MaxSteps -> 1000, DisplayFunction -> Identity,
  SaveLastPoint -> True];

LastPoint
0.529277
```

Note that one could get at this number without all the machinery of Visual-DSolve. The built-in solver can be used, though one must use the appropriate syntax for NDSolve as follows. We start with x[t], use the replacement rule to get an interpolating function, and then replace t with 20 to get the number we want.

```
x[t] /. NDSolve[{x'[t] + t Cos[Pi t]^2 x[t] == t Cos[t], x[0] == 0},
           x[t], {t, 0, 20}, MaxSteps -> 1000] /. t -> 20
{0.529277}
```

Now we switch to Runge–Kutta, using 200 steps (800 function evaluations) for comparison. The error at $t = 20$ is quite large.

```
VisualDSolve[x'[t] + t Cos[Pi t]^2 x[t] == t Cos[t],
  {t, 0, 20}, {x, -10, 10}, InitialValues -> {0, 0},
  DisplayFunction -> Identity, SaveLastPoint -> True,
  Method -> RungeKutta4, RKSteps -> 200];

LastPoint
0.6035815163580781
```

Chapter 2

Auxiliary Functions

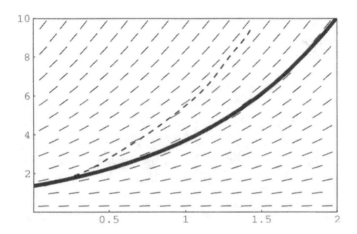

Overview

This chapter discusses some auxiliary functions that can be used with VisualDSolve. FreehandAttempt sets up a situation where a student can be asked to sketch a curve on the basis of the direction field, and then compare the guess to the real solution. PhaseLine draws an image of a 1-dimensional phase line for a single autonomous differential equation. ResidualPlot produces a graph of the result when a potential solution is substituted into a differential equation.

FUNCTIONS DISCUSSED: FreehandAttempt, GetPts, PhaseLine, ResidualPlot, ToSystem.

2.1 ▪ FreehandAttempt

Trying to use a direction field to get an idea of where a solution goes is not as easy as it sounds. For some examples it might be straightforward but for complicated situations it can be very difficult to guess a solution from the field of lines. The FreehandAttempt function allows a student to make an attempt and then compare it with the real solution. The first step is to generate a direction field by itself; we saw in section 1.7 how this can be done with VisualDSolve.

```
Needs["VisualDSolve`"];

VisualDSolve[x'[t] == x[t], {t, 0, 2}, {x, 0, 10},
  DirectionField -> True];
```

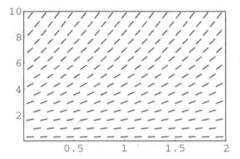

Now, by selecting a graphic, holding down the Command key (on a Macintosh; use the Control key if using an IBM-compatible), and clicking and dragging the mouse through the graphic, one can generate a set of points. After releasing the mouse and key, one can use the copy menu item to copy these items to the clipboard. They can then be pasted into a cell as follows, which will give them the name guess.

```
guess = {{-0.002995, 1.3575}, {0.029748, 1.3575}, {0.062492, 1.3575},
  {0.095235, 1.4238}, {0.12798, 1.49002}, {0.160723, 1.62247},
  {0.19347, 1.6887}, {0.22621, 1.7492}, {0.25894, 1.8211},
  {0.29169, 1.9536}, {0.32444, 2.0198}, {0.35718, 2.1522},
  {0.38992, 2.2847}, {0.42267, 2.4171}, {0.45541, 2.6158},
  {0.48816, 2.7483}, {0.52090, 2.9469}, {0.55364, 3.0794},
  {0.58639, 3.2118}, {0.61913, 3.4105}, {0.66006, 3.6092},
  {0.70917, 3.8079}, {0.74192, 4.0065}, {0.77466, 4.1390},
  {0.80741, 4.3377}, {0.84834, 4.6026}, {0.88926, 4.8012},
  {0.93019, 5.0661}, {0.97112, 5.2648}, {1.0038, 5.5297},
  {1.0366, 5.7284}, {1.0693, 5.9271}, {1.1021, 6.1920},
  {1.1348, 6.4569}, {1.1675, 6.7880}, {1.2003, 7.0526},
  {1.2248, 7.3178}, {1.2494, 7.5827}, {1.2821, 7.8476},
  {1.2985, 8.1125}, {1.3231, 8.3774}, {1.3476, 8.6423},
  {1.3804, 8.8410}, {1.3968, 9.1059}, {1.4213, 9.3708},
  {1.4377, 9.6357}, {1.4541, 9.9006}};
```

Here is what the screen might look like after an attempt with the mouse.

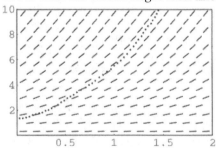

Now we can use `FreehandAttempt` to see how the guess (dashed) stacks up with the actual solution (thick) emanating from the same starting point as the guess.

```
FreehandAttempt[x'[t] == x[t], {t, 0, 2}, {x, 0, 10}, guess,
    DirectionField -> True, FieldThickness -> AbsoluteThickness[0.5]];
```

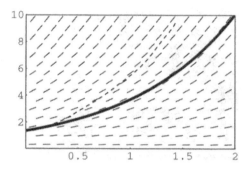

The dashed lines are obtained by an interpolation function that passes through every third point of the data in `guess`. If you wish to interpolate through every single point, set `Thinning` to `False` in the options to `FreehandAttempt`.

Even for this simple example, it is hard to come very close to the true solution. Here is a more complicated example.

```
VisualDSolve[x'[t] == x[t] - t Cos[t],
    {t, -5, 5}, {x, -4, 4},
    DirectionField -> True,
    FieldMeshSize -> {25, 20}];
```

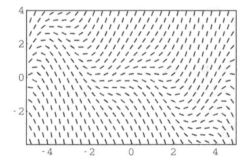

Now we can again use the Command (or Control) key and the mouse to trace out an attempt at a solution.

```
guess = {{-4.8512, -0.42737}, {-4.772, -0.29062},
  {-4.719183, -0.153873}, {-4.6927, 0.01706}, {-4.613536, 0.188009},
  {-4.481477, 0.35895}, {-4.37583, 0.529891}, {-4.217359, 0.700832},
  {-4.032476, 0.803396}, {-3.79477, 0.974337}, {-3.319358, 1.008525},
  {-3.21371, 0.905961}, {-3.081651, 0.803396}, {-2.949592, 0.598267},
  {-2.817533, 0.427326}, {-2.685474, 0.153821}, {-2.527004, -0.051308}
  {-2.421356, -0.290625}, {-2.289297, -0.529943},
  {-2.210062, -0.700884}, {-2.104415, -0.940201},
  {-1.972356, -1.179518}, {-1.919532, -1.350459},
  {-1.840297, -1.555588}, {-1.734649, -1.794905},
  {-1.655414, -1.931658}, {-1.549767, -2.136787},
  {-0.968707, -2.854739}, {-0.810236, -3.059868},
  {-0.678177, -3.19662}, {-0.546118, -3.333373},
  {-0.466883, -3.470126}, {-0.361236, -3.606879}};
```

```
FreehandAttempt[x'[t] == x[t] - t Cos[t],
  {t, -5, 5}, {x, -4, 4}, guess];
```

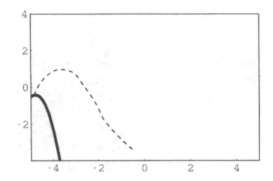

This one is quite difficult, because the solutions diverge as *t* increases. If the direction of time travel is reversed (this can be done by considering $x' = -x - \cos t$) then the solutions converge, and the exercise is significantly easier!

2.2 ▪ PhaseLine

PhaseLine produces a graphical view of a first-order autonomous differential equation. It contains the same information as a direction field only in more compact form. A solution to a first-order equation can be thought of as a particle moving along a line. In the autonomous case, the velocity is determined by the position and is independent of time. In particular, this means that a solution cannot double back over itself. Thus we can divide the line into regions where the particle is moving right and regions where the particle is moving left. PhaseLine represents the flow of the solution through each point on the line as an arrow where the length of the arrow represents how fast the flow is.

```
PhaseLine[x'[t] == Cos[x[t]], x[t], {x, -2 Pi, 2 Pi}];
```

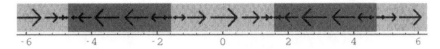

The darker background is where the flow is left and the light gray background is where the flow is right. The equilibrium points occur at the boundaries of the gray regions.

An alternative to the arrows that looks better in the regions where the arrows are too small to see is given by the `FlowField` option. This replaces the arrow with a shape whose thickness increases in the direction of the flow.

```
PhaseLine[x'[t] == Cos[x[t]], x[t], {x, -2 Pi, 2 Pi},
    FlowField -> True];
```

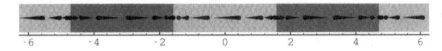

There are several options to `PhaseLine`.

```
Options[PhaseLine]
{FieldColor -> GrayLevel[0], NumberFish -> 10,
    FieldThickness -> Thickness[0.004],
    AspectRatio -> 0.1, TimeScale -> 10, FlowField -> False,
    FieldLogScale -> 1, FlowThickness -> 1}
```

`FieldLogScale` controls the length of the arrows or fish; `FlowThickness` controls the thickness of the fish. These are explained more fully in the context of vector fields and flow fields, discussed in detail in section 4.3. The two options to control the size of the arrows or fish-shaped objects are `TimeScale` and `NumberFish`. `TimeScale` represents the length of time over which the equation is solved for each initial condition (the initial conditions are determined internally). `NumberFish` determines how many arrows or fish will be used to represent the solution over the period of time specified by `TimeScale`. Thus each fish or vector represents a period of time approximately equal to `TimeScale/NumberFish`.

```
PhaseLine[x'[t] == Cos[x[t]], x[t], {x, -2 Pi, 2 Pi},
    FlowField -> True, TimeScale -> 2, NumberFish -> 1];
```

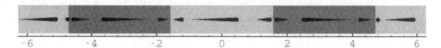

As a final example for `PhaseLine` consider the logistics model of population growth with constant harvesting. We leave the harvesting coefficient, c, as an unknown in defining the equation so that we may study what happens as the harvesting rate varies.

```
harvestEqn = (x'[t] == x[t] (3 - x[t]) - c);
```

We then make a movie to show the effect of increasing the harvesting coefficient from 0 to 3 in this example.

```
Do[PhaseLine[harvestEqn, x[t], {x, -1, 5},
    TimeScale -> 3, FlowField -> True],
    {c, 0, 3, 0.3}]
```

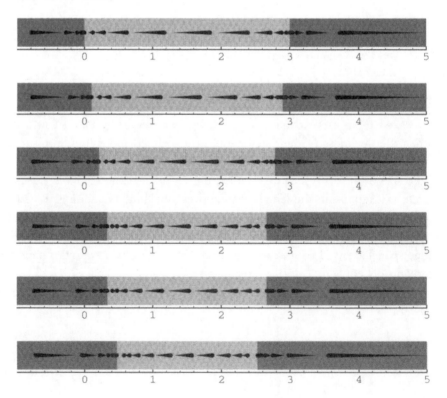

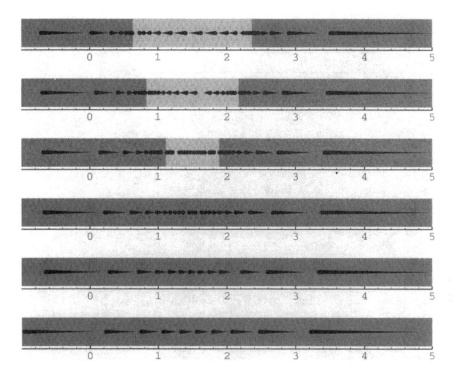

 With no harvesting the logistics model predicts the population will achieve an equilibrium state as long as the population is nonzero. As we increase harvesting we see that the population will still achieve an equilibrium but only so long as the population is not too small initially. At some point we see that an equilibrium is no longer present and that the population dies out for any initial value.

2.3 ResidualPlot

Because the output of NDSolve is an interpolating function, we can differentiate it and plug it back into the differential equation to see how well the equation is satisfied. The VisualDSolve package includes a function, ResidualPlot, that will do this automatically. First we must generate a solution, which we do by turning SaveSolution on. The default here is to save the solution in a global variable called Solution; but we can send a string via the SolutionName option so that a name of our choice is used, and in that case SaveSolution need not be mentioned.

```
VisualDSolve[x'[t] == x[t], {t, 0, 1}, {x, 0, 3},
    InitialValues -> {0, 1}, SolutionName -> "ourSoln"];
```

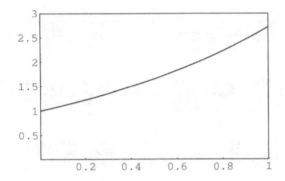

Now we check the solution. Since `ResidualPlot` plots the difference (known as the residual), a perfect solution corresponds to the zero function.

```
ResidualPlot[x'[t] == x[t], ourSoln, x[t], {t, 0, 1}];
```

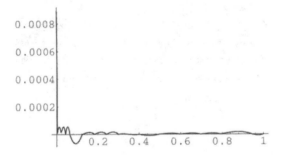

It often happens that the error near the initial value is quite large; this is because of `NDSolve`'s algorithm and how it adjusts step sizes and incorporates derivative data into the interpolation function as it learns it. Therefore one will probably want to restrict the plot range. The following command does that, and also adds some lines to indicate the *t*-values from the interpolation function's data points. `GetPtsFull` gets the points of an interpolation function (see section 2.4).

```
ResidualPlot[x'[t] == x[t], ourSoln, x[t], {t, 0, 1},
    PlotRange -> {0, 0.0001},
    Epilog -> {GrayLevel[0.5], AbsoluteThickness[0.5],
        Map[Line[{{#, 0}, {#, 0.0001}}] &,
            First /@ GetPtsFull[ourSoln]]}];
```

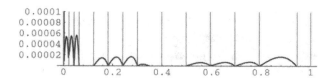

The vertical lines show us that the interpolating function best fits the equation in the vicinity of the interpolating points. The size of the error is comforting, though it does not prove anything about the distance between the numerical solution and the true solution. It simply says that the numerical solver has found a function that fits the equation pretty well. It will often be more convenient to make a logarithmic (base-10) plot, which is done as follows; this avoids the plot range issue.

```
ResidualPlot[x'[t] == x[t], ourSoln, x[t], {t, 0, 1},
    AxesOrigin -> {0, -8}, PlotType -> Logarithmic];
```

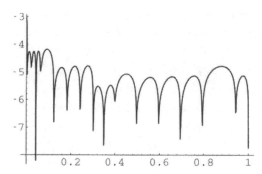

`ResidualPlot` will work for any function as the second argument, so we can use it to check symbolic solutions as well. Of course, if this works as hoped then the result is just a plot of the 0 function.

```
ResidualPlot[x'[t] == Cos[t] - x[t], C/E^t + (Cos[t] + Sin[t])/2,
    x[t], {t, 0, 10}, AxesOrigin -> {0, -1}, PlotRange -> {-1, 1}]
```

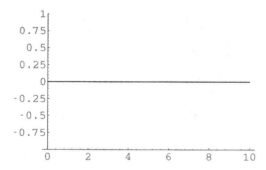

ResidualPlot allows us to get a better understanding of some of the precision-control options to NDSolve, which can be used by the solving functions of the VisualDSolve package as well. Three important options are WorkingPrecision, PrecisionGoal, and AccuracyGoal. WorkingPrecision controls the precision to which all the internal numerical computations will be done. The default is the machine precision of the computer being used. You can see what it is on your machine as follows. On a PowerMac it is 16 digits.

```
$MachinePrecision
```

```
16
```

The PrecisionGoal controls the local error that the numerical solver aims for at each step, in relative terms. The default is 6, which means that the numerical algorithm works until it feels it is correct to 6 significant digits. AccuracyGoal is similar, but applies to absolute error, as opposed to relative error. So the default of 6 for PrecisionGoal means that the method is working to achieve 6 digits of accuracy to the right of the decimal point independent of the size of the solution.

Now, if we set a PrecisionGoal of 15 we must increase the working precision to something at least that large. The default of 16 is not sufficient, since roundoff errors would grow very quickly; thus we use 25, which should suffice. Note also that one must be careful to include no approximate reals, such as 0.15, in the equation or initial conditions. An approximate real has only machine precision, and will contaminate the attempt to use 25 digits of working precision. Such input should be written as 15/100.

```
VisualDSolve[x'[t] == x[t], {t, 0, 1}, {x, 0, 3},
    InitialValues -> {0, 1}, WorkingPrecision -> 25,
    PrecisionGoal -> 15, SaveSolution -> True,
    DisplayFunction -> Identity];
```

```
ResidualPlot[x'[t] == x[t], Solution, x[t], {t, 0, 1},
    PlotType -> Logarithmic, AxesOrigin -> {0, -8}];
```

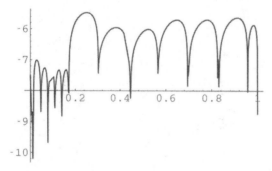

Unfortunately, a bug in version 2.2's `NDSolve` causes it to not work exactly as advertised. We can see the problem here in that the error has not shrunk to anywhere near 10^{-15}. This bug has been fixed in version 3.0 of *Mathematica*. Note that high-precision computations can take a long time. It turns out, in version 2.2, we get better results if we set both `PrecisionGoal` and `AccuracyGoal` to 15.

```
VisualDSolve[x'[t] == x[t], {t, 0, 1}, {x, 0, 3},
   InitialValues -> {0, 1}, WorkingPrecision -> 25,
   AccuracyGoal -> 15, PrecisionGoal -> 15,
   SaveSolution -> True, MaxSteps -> 4000,
   DisplayFunction -> Identity];
```

```
ResidualPlot[x'[t] == x[t], Solution, x[t], {t, 0, 1},
   PlotType -> Logarithmic, PlotRange -> {-16, -12}];
```

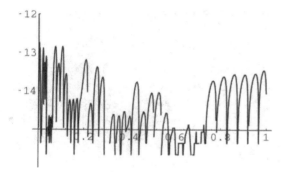

Good. This has given us a much improved error situation, with an interpolating function that fits the equation with about 14 digits of accuracy.

The preceding discussion is of more than passing interest. We have not addressed the very important issue of propagation error. For sensitive equations, a very small local error can propagate to quite large errors. Ignoring this sort of error can lead to solutions that look right, but are in fact very wrong. *Mathematica*'s high-precision capabilities allow us to detect such bad solutions, and also to compute solutions that, as far as we can tell, are correct. These topics are discussed more fully in section 15.5 (see also [KW]). But we mention here one interesting point. Suppose we obtain an interpolating function, $x(t)$, as the solution to a complicated initial-value problem, perhaps of the form $f(x'', x', x, t) = 0$, let the residual be $r(t)$. While it is true that a residual close to 0 does not necessarily mean that $x(t)$ is equally close to the true solution, note that $x(t)$ is indeed a perfect solution to the perturbed equation $f(x'', x', x, t) = r(t)$. If the residual is small, then it is quite possible that this perturbed equation and the original equation are indistinguishable from a physical perspective.

2.4 ▮ **GetPts**

GetPts and GetPtsFull are utilities that take an interpolating function and return the set of interpolating points.

```
VisualDSolve[y'[t] == y[t], {t, 0, 1}, {y, 0, 3},
  InitialValues -> {0, 1}, SaveSolution -> True,
  DisplayFunction -> Identity];
```

GetPts returns the *y*-values of the interpolating function.

GetPts[Solution]

```
{1., 1.0009, 1.00179, 1.02263, 1.0439, 1.06562, 1.13066,
  1.19968, 1.27291, 1.35061, 1.49043, 1.64473, 1.815,
  2.00289, 2.21024, 2.57409, 2.71829}
```

GetPtsFull returns the complete set of interpolation points.

GetPtsFull[Solution]

```
{{0., 1.}, {0.000894427, 1.0009}, {0.00178885, 1.00179},
  {0.0223764, 1.02263}, {0.042964, 1.0439}, {0.0635516, 1.06562},
  {0.122802, 1.13066}, {0.182053, 1.19968}, {0.241303, 1.27291},
  {0.300553, 1.35061}, {0.399063, 1.49043}, {0.497572, 1.64473},
  {0.596081, 1.815}, {0.69459, 2.00289}, {0.793099, 2.21024},
  {0.945494, 2.57409}, {1., 2.71829}}
```

This shows us that the numerical algorithm took 16 steps to solve the differential equation. This list can be useful for some specialized work involving the solution. An example is given in section 15.1.

2.5 ▮ **ToSystem**

ToSystem is primarily for internal use: it allows SecondOrderPlot to transform a second-order equation or system to a system of first-order equations. Its default is to use special auxiliary variables that vary from call to call.

```
ToSystem[{x"[t] + x[t] == t, y"[t] + x[t] == t},
  {x[t], y[t]}, t]

{x'[t] == x$1[t], y'[t] == y$2[t], x$1'[t] == t - x[t],
  y$2'[t] == t - x[t]}
```

However, the user who wishes to transform a second-order system in this way can specify the auxiliary variables as follows.

```
ToSystem[{x"[t] + x[t] == y'[t] t, y"[t] - 13 x[t] + y'[t] == t},
    {x[t], y[t]}, t, AuxiliaryVariables -> {u, v}] //TableForm

x'[t] == u[t]
y'[t] == v[t]
u'[t] == t v[t] - x[t]
v'[t] == t - v[t] + 13 x[t]
```

2.6 ■ ColorParametricPlot

`ColorParametricPlot` is a fancy version of `ParametricPlot` (see [Wag1] for a discussion of this function.) It produces a parametric plot of two functions where the direction of motion is indicated by changing colors. Its main use is as a setting to the `ParametricPlotFunction` option to `PhasePlot`; see page 73.

```
ColorParametricPlot[{Cos[4t]Cos[5t], Sin[t] Cos[3t]},
    {t ,0, 2 Pi}, PlotPoints -> 200, GrayShading -> True,
    PlotStyle -> {AbsoluteThickness[2]}];
```

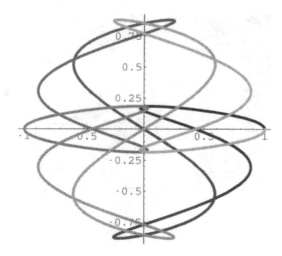

2.7 ■ FlowParametricPlot

`FlowParametricPlot` is another fancy version of `ParametricPlot`. It is written for internal use as an option to `ParametricPlotFunction` in the `VisualDSolve` package. Thus it returns only a `Graphics` object and you must use `Show` to display the image. In `FlowParametricPlot` the direction of motion is indicated by increasing the thickness of the curve. The curve is broken up into a number of pieces, controlled by the `NumberFish` option. Each fish-like object is made up of a number of polygons, controlled by the `Segments` option. Each fish-shape represents the same length of time, so it is very easy to see where the trajectory is fast or slow. Like

ColorParametricPlot, FlowParametricPlot can be used as a setting to the
ParametricPlotFunction option to PhasePlot; see page 73.

```
Show[FlowParametricPlot[{Sin[Sqrt[t]], Cos[t] Exp[-0.2 t]},
    {t, 0, 6 Pi}, {x, -1.1, 1.1}, {y, -1.1, 1.1},
    NumberFish -> 25, Segments -> 15,
    AspectRatio -> 1, FlowColor -> GrayLevel, Axes -> True],
    PlotRange -> {{-1.1, 1.1}, {-1.1, 1.1}}];
```

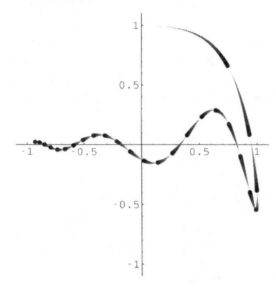

Chapter 3

SystemSolutionPlot

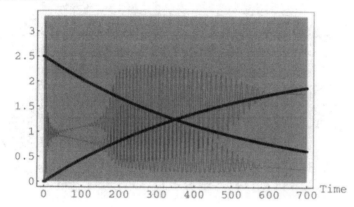

Overview

SystemSolutionPlot plots the graphs of the solutions to a system of first-order differential equations. It is often true that, for a system involving equations for dx/dt and dy/dt, a phase plot (x vs. y) is more informative than individual plots of x and y as functions of t. Yet the individual plots can also be important. They show more clearly how the variables change as a function of t. SystemSolutionPlot allows the user to see the solutions either combined in a single plot or in separate plots.

FUNCTIONS DISCUSSED: SystemSolutionPlot.

3.1 ■ Basic Usage

In its basic form SystemSolutionPlot takes a system of two or more equations, a list of the dependent variables, a t-iterator, and a set of initial values. The output consists of a single image containing the graphs of each variable against t.

```
Needs["VisualDSolve`"];

SystemSolutionPlot[{x'[t] == y[t], y'[t] == -x[t], z'[t] == 0.5},
    {x[t], y[t], z[t]}, {t, 0, 2 Pi}, InitialValues -> {1, 0, 1.5}];
```

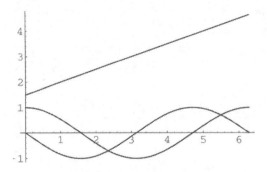

Just as with `VisualDSolve`, we can almost always save some time by using `FastPlotting` (as explained on page 14). This example is a simple one, but still `FastPlotting` cuts the time by a factor of two.

```
SystemSolutionPlot[{x'[t] == y[t], y'[t] == -x[t], z'[t] == 0.5},
   {x[t], y[t], z[t]}, {t, 0, 2 Pi},
   InitialValues -> {1, 0, 1.5}, FastPlotting -> True];
```

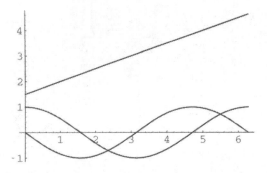

We can use more initial values, but then quite complicated plots can arise, and some styles will have to be set to distinguish them.

```
SystemSolutionPlot[{x'[t] == y[t], y'[t] == -x[t], z'[t] == 0.5},
   {x[t], y[t], z[t]}, {t, 0, 2 Pi}, FastPlotting -> True,
   InitialValues -> {{1, 0, 1.5}, {2, 2, 2}}];
```

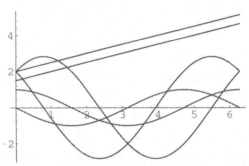

Alternatively, we can plot only some of the variables. This is done via the `PlotVariables` option. The default is to show everything on one image; if there are three variables, then this corresponds to a setting of `{{x, y, z}}`, which is interpreted as: show one image with all three functions plotted. The following example creates a single image with plots of $x(t)$ and $z(t)$. We also add specific t-values to the front of the initial value list, which allows us to work both left and right from the initial condition.

```
SystemSolutionPlot[{x'[t] == y[t], y'[t] == -x[t], z'[t] == 0.5},
  {x[t], y[t], z[t]}, {t, 0, 2 Pi},
  InitialValues -> {3, 1, 0, 1.5}, PlotVariables -> {{x, z}},
  PlotLabel -> "x (lower) and z vs. t"];
```

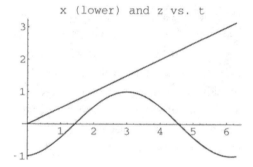

To get several plots, we use more lists in the `PlotVariables` setting. The `PlotLabels` option provides a way of attaching different labels to the plots.

```
SystemSolutionPlot[{x'[t] == y[t], y'[t] == -x[t], z'[t] == 0.5},
  {x[t], y[t], z[t]}, {t, 0, 2 Pi},
  InitialValues -> {{3, 1, 0, 1.5}, {3, 2, 2, 2}},
  PlotVariables -> {{x}, {z}},
  PlotLabels -> {"x vs. t", "z vs. t"}];
```

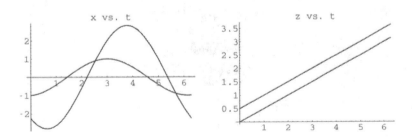

The `PlotRanges` option allows us to send different plotting ranges to the different plots.

```
SystemSolutionPlot[{x'[t] == y[t], y'[t] == -x[t], z'[t] == 1},
  {x[t], y[t], z[t]}, {t, 0, 2 Pi},
  InitialValues -> {1, 0, 1.5}, PlotVariables -> {{x}, {z}},
  PlotRanges -> {{-1.2, 2}, All}];
```

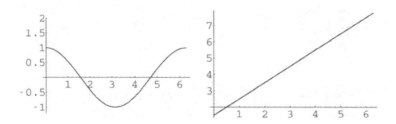

And we can vary the initial conditions. Here is an example that shows what happens to a driven pendulum with a variety of initial conditions. The equations come from the second-order equation $x'' = \cos t - 0.1x' - \sin x$, where x represents the pendulum's angle. The initial x-value is 0, but the initial velocity varies from 1.85 to 2.1. The solutions settle down, but it is difficult to predict which periodic solution they will converge to. It turns out that the basins of attraction form a very complicated topological set called the Lakes of Wada; this example is discussed in much more detail in Chapter 16. A color version of this plot, using the Rainbow option discussed in the next section, is in color plate 4.

```
SystemSolutionPlot[
  {x'[t] == v[t], v'[t] == Cos[t] - Sin[x[t]] - 0.1 v[t]},
  {x[t], v[t]}, {t, 0, 200},
  InitialValues -> Table[{0, v}, {v, 1.85, 2.1, 0.025}],
  PlotVariables -> {{x}}, MaxSteps -> 4000,
  PlotStyle -> AbsoluteThickness[0.5],
  PlotRange -> {-35, 23}, PlotPoints -> 200];
```

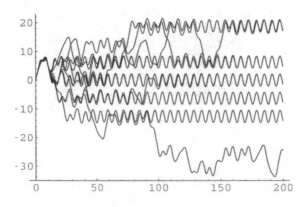

3.2 ▪ **Stylish Plots**

We can use the following options from VisualDSolve: PlotStyle, SaveSolution, and SolutionName, as well as the usual Graphics options (such as Background or AxesLabel). But the PlotStyle option can be used provided only a single image has been requested. Then one should send a list of styles: the first applies to the first plotting variable, the second to the second plotting variable, and so on.

```
SystemSolutionPlot[{x'[t] == y[t], y'[t] == -x[t], z'[t] == 1},
  {x[t], y[t], z[t]}, {t, 0, 2 Pi},
  PlotStyle -> {{AbsoluteThickness[0.5]},
                {AbsoluteThickness[1.5]},
                {AbsoluteThickness[3]}},
  InitialValues -> {1, 0, 1.2},
  PlotRange -> {{-0.1, 2 Pi}, {-1.05, 2}},
  Background -> GrayLevel[0.8]];
```

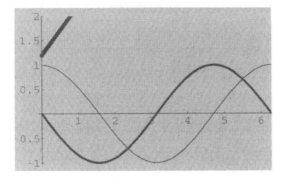

A single setting for PlotStyle (or a single list) causes the attribute(s) to apply to all plots.

```
SystemSolutionPlot[{x'[t] == y[t], y'[t] == -x[t], z'[t] == 1},
  {x[t], y[t], z[t]}, {t, 0, 2 Pi},
  PlotStyle -> AbsoluteThickness[3],
  InitialValues -> {1, 0, 1.2},
  PlotRange -> {{-0.1, 2 Pi}, {-1.05, 2}}];
```

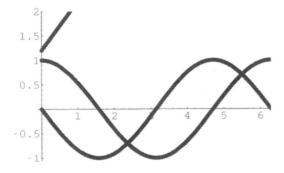

Recall that `Rainbow` is used to request different colors for different initial values. That too works only if a single image has been requested. Then setting `Rainbow` to `True` causes different hues to be used for different initial values, even if, say, plots of $x(t)$ and $y(t)$ are combined in one image. If the user attempts to use `Rainbow` or `PlotStyle` when more than one image is requested, a warning message results. In such a case the user will have to resort to saving the solutions via the `SaveSolution` option and plotting them separately, with the appropriate plot styles.

Here is an interesting example that stems from a type of chemical reaction called an *autocatalator*; it is discussed in more detail in [BCB, pp. 292–293]. As with many complicated examples, this one requires lots of memory. If you are working on a computer with a modest amount of memory, the simplest solution is to reduce `tmax` from 700 to, say, 200. In this example we illustrate the use of `Prolog` to create an image with a background that enhances the colors of the plot. While this effect could be achieved by setting the `Background` option to a shade of gray, that would provide a background for everything, including the regions left and below the axes. We instead use the graphics capabilities of *Mathematica* to add a rectangle and some lines in the prolog, moving the axes origin and setting the plot range appropriately. We use `Prolog` so that the curves are drawn on top of the gray background. Since we run t all the way out to 700, we must increase `MaxSteps`. The output can be found in color plate 5. Note also that one can use indexed variables, such as `x[1]`.

```
tmax = 700;
m1 = 700/40;
m2 = 3.3/30;
thin = Thickness[0.0004];
thick = Thickness[0.01];

SystemSolutionPlot[{
  x[1]'[t] == -0.002 x[1][t],
  x[2]'[t] ==    0.4 x[1][t] - 0.08 x[2][t] - x[2][t] * x[3][t]^2,
  x[3]'[t] ==       - x[3][t] + 0.08 x[2][t] + x[2][t] * x[3][t]^2,
  x[4]'[t] ==    0.005 x[3][t]},
  Table[x[i][t], {i, 1, 4}], {t, 0, tmax},
  InitialValues -> {2.5, 0, 0, 0},
  PlotStyle -> {{Green, thick}, {Yellow, thin}, {Red, thin},
                {Cyan, thick}},
  MaxSteps -> 4000, AxesOrigin -> {-m1, -m2},
  AxesLabel -> {Time, Concentrations},
  Prolog -> {{Thickness[0.002], Line[{{-m1, 3.3 + m2},
          {tmax + m1, 3.3 + m2}, {tmax + m1, -m2}}]],
          Gray, Rectangle[{0, 0}, {tmax, 3.3}]}},
  PlotRange -> {{-m1, tmax + m1}, {-m2, 3.3 + m2}},
  PlotPoints -> 200, Ticks -> {Automatic, Range[0, 3, 0.5]}}];
```

Concentrations

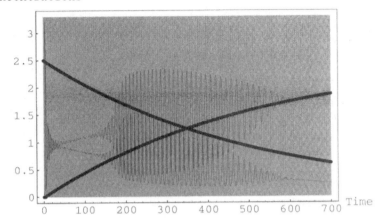

This is a tough one to tweak properly for smoothness. There are four inter-polating functions, each of which involves 3297 points (the number of steps used by the numerical algorithm to solve the system). If we use `FastPlotting`, then we are asking for over 12,000 plotting points in all, which not only requires an inordinate amount of memory, but is somewhat silly since two of the curves are quite smooth, and can be plotted with about 100 plotting points.

So, despite the general benefits of the `FastPlotting` option, it is best here to turn it off and use the default plotting mechanism with a `PlotPoints` setting of 200. For the two smooth curves, the 200 points will see nicely small angles, and the adaptive plotting routine will stop. For the bumpy curves enough adaption will take place to catch all the bumps. Because the default number of initial function evaluations for `Plot` (the `PlotPoints` option) is merely 25, some oscillations can be missed in a function with 30 or more oscillations. The reader might wish to redo the preceding example with the default settings (`FastPlotting` turned off and no special `PlotPoints` setting) and compare the results to the image shown.

If we wish to vary one of the coefficients, that is easily done by defining a function to generate the solution for a particular coefficient, but suppress the display. Then we can show three solutions at once, in this case observing that the highest setting of the parameter suppresses the oscillations beyond $t = 100$.

```
Autocatalator[a_] := SystemSolutionPlot[{
   x[1]'[t] == -0.002 x[1][t],
   x[2]'[t] ==    0.4 x[1][t] - a x[2][t] - x[2][t] * x[3][t]^2,
   x[3]'[t] ==      - x[3][t] + a x[2][t] + x[2][t] * x[3][t]^2,
   x[4]'[t] ==  0.005 x[3][t]},
   Table[x[i][t], {i, 1, 4}], {t, 0, 700},
   InitialValues -> {2.5, 0, 0, 0}, DisplayFunction -> Identity,
   PlotVariables -> {{x[2]}}, PlotRange -> {0, 3.5},
   PlotStyle -> AbsoluteThickness[a /.
       {0.08 -> 0.5, 0.10 -> 1.5, 0.14 -> 2.5}],
   PlotPoints -> 200,
   MaxSteps -> 4000, Ticks -> {Automatic, Range[0, 3, 0.5]}];

fo[s_] := FontForm[s, {"Courier", 8}];

Show[Map[Autocatalator, {0.08, 0.10, 0.14}],
   DisplayFunction -> $DisplayFunction,
   Epilog -> {
     Text[fo["a = 0.08"], {617, 1.84}, {-1, -1}],
     Text[fo["a = 0.10"], {617, 1.30}, {-1, -1}],
     Text[fo["a = 0.14"], {617, 0.90}, {-1, -1}]},
   AxesLabel -> {t, x[2]}];
```

3.3 ■ Using the Output

We can combine several plots into a single image by using a nice feature of the Rectangle command. Rectangle[p, q] represents a rectangle with lower-left corner p and upper-right corner q. But one can add a third argument, im. Rectangle[p, q, im] will place the image im in the aforementioned rectangle. Getting the aspect ratios right is crucial to getting a nice image. In the example that follows, we use the default aspect ratios for the plots of *x* and *y* against *t*, but for the phase plot (this package function is discussed in full detail in Chapter 4) we use an aspect ratio of 1.8/1.2, since we wish it to fit in a rectangle of those dimensions. We then display all three images by showing the `Graphics` object consisting of three rectangles, with the rectangle coordinates chosen to coordinate with the aspect ratios of the corresponding plots.

```
eqn = {x'[t] == y[t], y'[t] == 0.5 x[t] - 0.5 x[t]^3 - 0.015 y[t]};

{plotx, ploty} = SystemSolutionPlot[eqn, {x[t], y[t]}, {t, 0, 50},
    InitialValues -> {1, 0.54}, PlotVariables -> {{x}, {y}},
    FastPlotting -> True, DisplayFunction -> Identity];

plotxy = PhasePlot[eqn, {x[t], y[t]},
    {t, 0, 50}, {x, -2, 2}, {y, -1, 1},
    InitialValues -> {1, 0.54}, AspectRatio -> 1.8/1.2,
    AxesLabel -> {x, y}, FastPlotting -> True,
    DisplayFunction -> Identity];

Show[Graphics[{
    Rectangle[{0.4,  0}, {0.4 + 0.8 GoldenRatio,  0.8}, plotx],
    Rectangle[{0.4, -1}, {0.4 + 0.8 GoldenRatio, -0.2}, ploty],
    Rectangle[{-1,  -1}, {0.2, 0.8}, plotxy]}]];
```

See section 4.8 for a discussion of how to create a single three-dimensional image that shows all the images of the preceding example in a single three-dimensional plot.

Chapter 4

PhasePlot

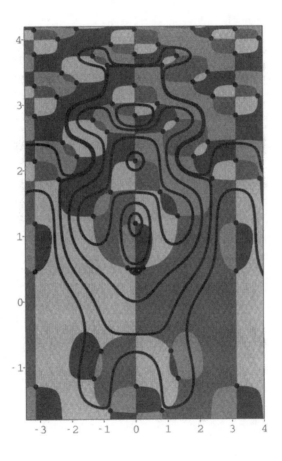

Overview

PhasePlot plots orbits for systems of ordinary differential equations. There are several complications, since such orbits can be high-dimensional objects and can be viewed in either two or three dimensions. As usual, there are lots of options to enhance the images one can produce. The most important ones are the use of arrows or fish to represent the flow, the display of the nullcline curves and shaded nullcline regions to enhance our understanding of the flow, and the computation and classification of the equilibrium points.

For some situations it will be more appropriate to use SystemSolutionPlot, which plots all the solutions as a function of the independent variable; see Chapter 3. And when dealing with second-order equations (or systems), there is no need to convert to a first-order system via auxiliary variables. This is all done automatically when one uses SecondOrderPlot (Chapter 5).

FUNCTIONS DISCUSSED: Equilibria, PhasePlot, PoincareSection.

4.1 ■ Basic Usage

PhasePlot produces diverse views of orbits for systems of ordinary differential equations. The basic usage is similar to VisualDSolve, with the initial values fed in as an option. Note that while an initial value to VisualDSolve consists of a *t-x* pair, here an initial value must have the form {t0, x0, y0} (with more entries for larger systems). However, since autonomous systems are so prevalent, we allow the t0 to be omitted, in which case it is set to the minimum *t*-value in the main *t*-iterator. Many of the options are the same as for VisualDSolve, and our first example illustrates the use of AxesLabel, and AspectRatio. The last is useful here (recall: Automatic means a scale on the axes that matches the true length of the axes) since this example is circular in nature, and the default aspect ratio, the reciprocal of the golden ratio, would give it an elliptical look. PhasePlot requires that the second argument be a list of the dependent variables.

```
Needs["VisualDSolve`"];

PhasePlot[{x'[t] == y[t], y'[t] == -x[t] - y[t]/10},
   {x[t], y[t]}, {t, 0, 5 Pi}, {x, -1.4, 1.4}, {y, -1.4, 1.4},
   InitialValues -> {0, 1}, ShowInitialValues -> True,
   InitialPointStyle -> AbsolutePointSize[4],
   AspectRatio -> Automatic, AxesLabel -> {x, y}];
```

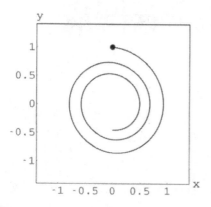

As with `VisualDSolve`, we can use several initial values.

```
PhasePlot[{x'[t] == y[t], y'[t] == -x[t] - y[t]/10},
   {x[t], y[t]}, {t, 0, 5 Pi}, {x, -1.4, 1.4}, {y, -1.4, 1.4},
   InitialValues -> {{0, 1}, {0, 1.2}},
   InitialPointStyle -> AbsolutePointSize[4],
   ShowInitialValues -> True, AspectRatio -> Automatic];
```

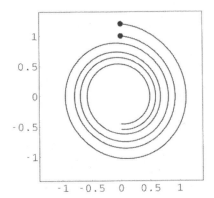

Here is a nonautonomous example that illustrates the fact that the initial *t*-value can be different than the smaller value in the *t*-iterator, which controls the plotting domain. Here the initial values are given in two types: the last two specify $t_0 = 1$, while the first relies on the default ($t_0 = 0$).

```
PhasePlot[{x'[t] == 1, y'[t] == t}, {x[t], y[t]},
   {t, 0, 2}, {x, -1.4, 2.5}, {y, 0, 3.6},
   InitialValues -> {{0, 1}, {1, 0, 1}, {1, 0, 2}},
   ShowInitialValues -> True, AspectRatio -> Automatic,
   InitialPointStyle -> AbsolutePointSize[4]];
```

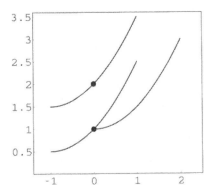

4.2 ■ Controlling the Style of the Orbits

When it is given a list of graphics primitives for each orbit, the PlotStyle option causes the orbits to be defined in the corresponding styles. The color output of the following is in color plate 6.

```
PhasePlot[{x'[t] == y[t], y'[t] == -x[t] - y[t]/10},
    {x[t], y[t]}, {t, 0, 5 Pi}, {x, -1.25, 1.25}, {y, -1.25, 1.25},
    InitialValues -> {{0, 1}, {0, 1.1}, {0, 1.2}},
    PlotStyle -> {{Red}, {Green}, {Blue}},
    AspectRatio -> Automatic];
```

Sometimes showing the initial values is sufficient in order to allow a quick determination of the direction of the orbits. The ShowInitialValues option accomplishes this. Another useful way to indicate the direction of the orbits is to ask for a little arrow to be placed on each one.

```
PhasePlot[{x'[t] == y[t], y'[t] == -x[t] - y[t]/10},
    {x[t], y[t]}, {t, 0, 5 Pi}, {x, -1.25, 1.25},
    {y, -1.25, 1.25},
    InitialValues -> {{0, 1}, {0, 1.1}, {0, 1.2}},
    DirectionArrow -> True, AspectRatio -> Automatic];
```

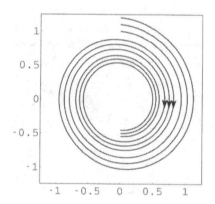

Placing direction arrows is difficult to do in a way that will be satisfactory in all examples. The main problem is that the arrows can crash into each other if the orbits are close. It is much clearer to color the images so that the changing colors indicate the direction of motion. Recall that these plots are all parametric plots of two functions, $x(t)$ and $y(t)$. The desired coloring can be achieved by using a fancier version of ParametricPlot called ColorParametricPlot, which is a function in the VisualDSolve package (see [Wag1] for a discussion of this function). To call this function, just set ParametricPlotFunction to ColorParametricPlot. We use colors going from cold (blue) to hot (red) to indicate the increasing direction. In the next example we keep the arrows to further emphasize the direction. The use of a gray window shade enhances color perception. The color output is in color plate 7.

```
PhasePlot[{x'[t] == y[t], y'[t] == -x[t] - y[t]/10},
    {x[t], y[t]}, {t, 0, 5 Pi}, {x, -1.3, 1.3}, {y, -1.3, 1.3},
    InitialValues -> {{0, 1}, {0, 1.1}, {0, 1.2}},
    ParametricPlotFunction -> ColorParametricPlot,
    PlotStyle -> AbsoluteThickness[2], AspectRatio -> Automatic,
    WindowShade -> GrayLevel[0.5], DirectionArrow -> True];
```

For black-and-white printing, the user will want to set the GrayShading option to True to get orbits that vary in shades of gray from black to white (really, very light gray) as *t* increases.

```
PhasePlot[{x'[t] == y[t], y'[t] == -x[t] - y[t]/10},
    {x[t], y[t]}, {t, 0, 5 Pi},
    {x, -1.25, 1.25}, {y, -1.25, 1.25},
    InitialValues -> {{0, 1}, {0, 1.1}, {0, 1.2}},
    ParametricPlotFunction -> ColorParametricPlot,
    PlotStyle -> AbsoluteThickness[3], AspectRatio -> Automatic,
    DirectionArrow -> True, GrayShading -> True];
```

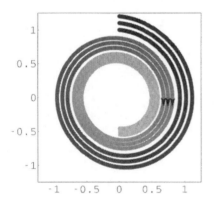

One can control the thickness by using PlotStyle, which passes its settings to ColorParametricPlot, but does not affect the coloring of the curve.

```
PhasePlot[{x'[t] == y[t], y'[t] == -x[t] - y[t]/10},
    {x[t], y[t]}, {t, 0, 5 Pi},
    {x, -1.25, 1.25}, {y, -1.25, 1.25},
    InitialValues -> {{0, 1}, {0, 1.2}},
    ParametricPlotFunction -> ColorParametricPlot,
    PlotStyle -> {{AbsoluteThickness[3]}, {AbsoluteThickness[1]}},
    AspectRatio -> Automatic, GrayShading -> True];
```

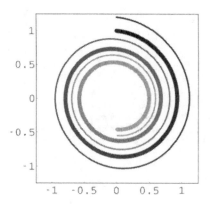

There is another useful way that `ColorParametricPlot` can shade its curves: it can use the speed of motion. Again, recall that an image of an orbit tells us only the curve on which a particle will travel if it adheres to the defining differential equation; it says nothing about the speed of the particle as it moves. `ColorParametricPlot` accepts a `Color` option that can be set to `Speed`, which is then used, in a normalized fashion, to color the curve using varying hues (from blue (slow) to red (fast)), or shades of gray (from black (slow) to light gray (fast)) if `GrayShading` is set to `True`.

```
PhasePlot[{x'[t] == y[t], y'[t] == -x[t] - y[t]/10},
    {x[t], y[t]}, {t, 0, 5 Pi},
    {x, -1.25, 1.25}, {y, -1.25, 1.25},
    InitialValues -> {0, 1},
    ParametricPlotFunction -> ColorParametricPlot,
    Color -> Speed, GrayShading -> True,
    PlotStyle -> AbsoluteThickness[3],
    AspectRatio -> Automatic, DirectionArrow -> True];
```

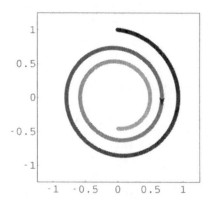

4.3 ■ Direction Fields and Flow Fields

Systems of two first-order autonomous differential equations provide the most striking images of the geometric nature of differential equations. In this situation the differential equation describes a vector field in the plane and there is a unique orbit through each point except where the associated vector field is singular.

The `VectorField` option to `PhasePlot` shows the vector field associated to a system of two first-order autonomous differential equations. The lengths of the vectors represent, in a scaled way, the actual lengths of the vectors in the field. `PhasePlot` will generate an error message if the equation is not autonomous or is not a system of two equations.

```
PhasePlot[{x'[t] == y[t], y'[t] == -x[t] - y[t]/10},
    {x[t], y[t]}, {t, 0, 6}, {x, -3, 3}, {y, -3, 3},
    InitialValues -> {0, 2}, VectorField -> True,
    AspectRatio -> Automatic, ShowInitialValues -> True,
    InitialPointStyle -> AbsolutePointSize[4]];
```

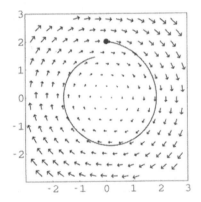

If the options are set to request a vector field without any superimposed orbits, then the *t*-iterator need not be included.

```
PhasePlot[{x'[t] == y[t], y'[t] == -x[t] - y[t]},
    {x[t], y[t]}, {x, -3, 3}, {y, -3, 3}, VectorField -> True];
```

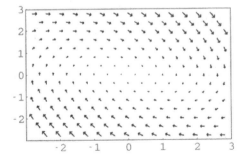

The vector-lengths can be controlled in two ways, via either `FieldLength` or `FieldLogScale`. `FieldLength` has a default of 0.8 which means that the longest vectors use up about 80% of the space to the adjacent vectors. With `FieldLength` greater than 1, the longer vectors might run into each other, but such a setting does sometimes give a better feel for the flow of the vector field.

```
PhasePlot[{x'[t] == y[t], y'[t] == -x[t] - y[t]},
   {x[t], y[t]}, {x, -3, 3}, {y, -3, 3},
   VectorField -> True, FieldLength -> 1.5,
   AspectRatio -> Automatic];
```

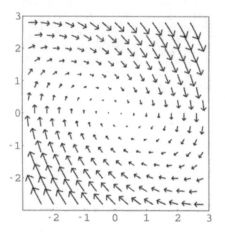

`FieldLogScale` provides a method for stretching out the shorter vectors in the plot without making the long vectors excessively long. This is especially useful for seeing the behavior of the vector field near equilibrium points. `FieldLogScale` has a default of 1 and must be between 1 and 10. By default, all vectors are scaled by the same factor so that the longest vector fits in the grid. Setting `FieldLogScale` to n further scales the vectors by $(1/n)$Log[10, *relative length*], so that for $n = 1$ no further scaling is done and for $n = 10$ all vectors are scaled to the same length. A very handy choice is `FieldLogScale -> Automatic` which causes a value to be automatically calculated so that the 1/4 longest vector in the field appears to be 1/4 the size of the longest vector. The next image compares an `Automatic` setting of `FieldLogScale` (right) to the default (left).

```
Show[GraphicsArray[{
   PhasePlot[{x'[t] == y[t] - x[t], y'[t] == 5 y[t] - x[t]^3 + 20 x[t]},
     {x[t], y[t]},  {x, -10, 10}, {y, -10, 10},
     AspectRatio -> Automatic,
     VectorField -> True, DisplayFunction -> Identity],
   PhasePlot[{x'[t] == y[t] - x[t], y'[t] == 5 y[t] - x[t]^3 + 20 x[t]},
     {x[t], y[t]}, {x, -10, 10}, {y, -10, 10},
     FieldLogScale -> Automatic, AspectRatio -> Automatic,
     VectorField -> True, DisplayFunction -> Identity]}],
 DisplayFunction -> $DisplayFunction];
```

FieldThickness accepts instructions for controlling the thickness of the vectors. FieldColor can be either a plain color such as Blue or LightSalmon or one of four special values: GrayLevel, Hue, RandomGrayLevel, or RandomHue. A setting of Hue colors the vectors from blue (shortest) to red (longest). A setting of GrayLevel uses shades of gray to color the vectors. RandomGrayLevel and RandomHue use random gray levels and hues, respectively, for each arrow. FieldMeshSize accepts an integer (or a pair of integers). The default is 15, which calls for a 15×15 mesh.

```
PhasePlot[{x'[t] == -Cos[y[t] + x[t]], y'[t] == Sin[x[t]^2]},
   {x[t], y[t]}, {x, -3, 3}, {y, 0, 4},
   VectorField -> True, FieldColor -> GrayLevel, FieldMeshSize -> 10,
   FieldLength -> 1.1, FieldThickness -> Thickness[0.008]];
```

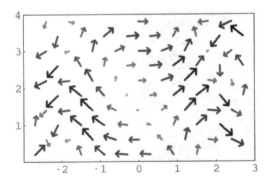

Here is a version in color, with `FieldLogScale` set to 10, thus causing all the arrows to have the same length. The color output is in color plate 8.

```
PhasePlot[{x'[t] == -Cos[y[t] + x[t]], y'[t] == Sin[x[t]^2]},
  {x[t], y[t]}, {x, -3, 3}, {y, 0, 4},
  VectorField -> True, FieldColor -> Hue,
  FieldThickness -> Thickness[0.006],
  FieldMeshSize -> 25, FieldLogScale -> 10, WindowShade -> Gray];
```

An alternative to drawing a vector field is to show the flow of the vector field by drawing many short trajectories. The main problem is finding a good way to visualize the direction of flow. The idea of using fish-shapes is due to Kaplan and Glass, whose book [KG] contains some of the history of ideas related to perception of flow. We call this the `FlowField` option, since it better shows the flowing nature of the underlying vector field. The color output is in color plate 9.

```
PhasePlot[{x'[t] == -Cos[y[t] + x[t]], y'[t] == Sin[x[t]^2]},
  {x[t], y[t]}, {x, -3, 3}, {y, 0, 4},
  FlowField -> True, FieldLogScale -> Automatic,
  FieldMeshSize -> 25, WindowShade -> Gray, FieldColor -> Hue];
```

Using `NDSolve` to calculate the flows would be very expensive and would make it difficult to control the lengths of the fish. We've opted to use a 3-term Taylor series method since we can get the curved trajectories using a single step and can control the length of the trajectories by calculating the sizes of the linear and quadratic terms separately. `FlowThickness` is used to scale the width of the fish; this option has a default of 1. `FieldLength`, `FieldColor`, and `FieldMeshSize` work as they do for the `VectorField` option. Here we generate a striking image by using random shades of gray on a black background.

```
PhasePlot[{x'[t] == y[t], y'[t] == -Sin[x[t]] - 0.1 y[t]},
  {x[t], y[t]}, {x, -Pi, 2 Pi}, {y, -3, 3},
  FlowField -> True, FieldMeshSize -> 30, FieldLength -> 2,
  WindowShade -> Black, FieldColor -> RandomGrayLevel];
```

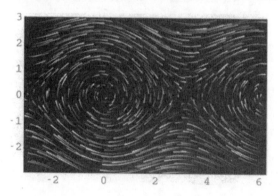

4.4 ■ FlowParametricPlot

A second option for `ParametricPlotFunction` is `FlowParametricPlot`. This version of `ParametricPlot` draws the curve as a series of pieces having increasing width in the direction of the flow.

```
PhasePlot[{x'[t] == -Cos[y[t] + x[t]], y'[t] == Sin[x[t]^2]},
    {x[t], y[t]}, {t, -3, 3}, {x, 1/2, 3}, {y, -1, 3},
    InitialValues -> {0, 3/2, 3/2},
    ParametricPlotFunction -> FlowParametricPlot];
```

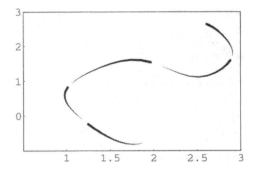

The number of pieces ("fish shapes") that each curve breaks up into is controlled by the `NumberFish` option. Each fish shape represents a fixed length of time, approximately `(tf - ti)/NumberFish` where the time iterator is `{t, ti, tf}`. This is helpful because it gives insight into the three-dimensional nature of parametric plots. This also makes life difficult as it can be difficult to decide on appropriate *t*-iterators and `NumberFish`. `Segments` controls how many points are used to plot each fish shape. In general a longer *t*-iterator or smaller value of `NumberFish` will require a larger value of `Segments` to get a smooth picture.

```
PhasePlot[{x'[t] == -Cos[y[t] + x[t]], y'[t] == Sin[x[t]^2]},
    {x[t], y[t]}, {t, -6, 6}, {x, 1/2, 3}, {y, -1, 3},
    InitialValues -> {0, 3/2, 3/2},
    ParametricPlotFunction -> FlowParametricPlot,
    NumberFish -> 40, Segments -> 5];
```

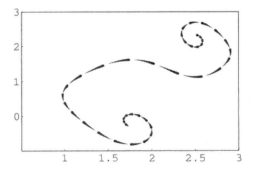

The appearance of the fish are controlled by `FlowThickness` and `FlowColor`. `FlowThickness` is a relative scaling factor with a default of 1.0; a setting of 2 results in fish that are twice as thick. `FlowColor` can be used to set the color of the fish.

`StayInWindow` is a very useful option in this context as it saves a lot of `Polygon` objects, thus saving on both time and memory consumption. One can also use random initial values to generate very effective phase portraits; one would set the initial values to, say, `Table[Random[Real, {x0, x1}], Random[Real, {y0, y1}], {30}]`.

```
PhasePlot[{x'[t] == -Cos[y[t] + x[t]], y'[t] == Sin[x[t]]},
   {x[t], y[t]}, {t, -4, 4}, {x, -4, 4}, {y, -3, 3},
   InitialValues -> Flatten[Table[
      {0, i, j}, {i, -3.5, 3.5}, {j, -2, 2}], 1],
   ParametricPlotFunction -> FlowParametricPlot,
   StayInWindow -> True,
   FlowThickness -> 0.75,
   WindowShade -> GrayLevel[0.2],
   FlowColor -> GrayLevel[0.8]];
```

The `Rainbow` option to `PhasePlot` also works with `FlowParametricPlot`. We use this option to make an instructive picture for the differential equation arising from the polar form $r' = \sin r$, $\theta' = 1$. It is easy to see that for $0 < r < \pi$ the curves are spiraling out; this follows from the fact that $r' > 0$. For $\pi < r < 2\pi$ the curves are spiraling in, because $r' < 0$. The color output is in color plate 10.

```
r = Sqrt[x[t]^2 + y[t]^2];
eqn = {x'[t] == Sin[r]/r x[t] - y[t],
       y'[t] == Sin[r]/r y[t] + x[t]};

PhasePlot[eqn, {x[t], y[t]}, {t, -3, 3}, {x, -9, 9}, {y, -9, 9},
   InitialValues -> Flatten[Table[{0, j Sin[i], j Cos[i]},
      {j, Pi/2, 5 Pi/2, Pi}, {i, Pi/5, 2 Pi, Pi/5}], 1],
   ParametricPlotFunction -> FlowParametricPlot,
   AspectRatio -> 1, FlowThickness -> 0.5,
   StayInWindow -> True, WindowShade -> Gray,
   Segments -> 30, Rainbow -> True, NumberFish -> 5];
```

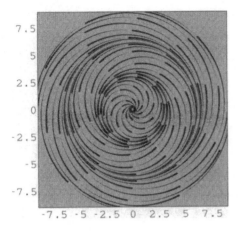

4.5 ■ Controlling the *t*-Domain

One important issue involving *t*-domains arises when one is interested in the limiting behavior of an orbit or orbits. As with `VisualDSolve`, we can add restrictions to the plotting interval to the initial value settings. More precisely, an initial value of the form `{t0, x0, y0, {tmin, tmax}}` causes the initial condition to be `x[t0]` = x0 and `y[t0]` = y0, and then shows the orbit from `tmin` to `tmax`. If the first `t0` is omitted, it is taken from the main *t*-iterator. Here is an example involving the van der Pol equation which we first show with *t* running from 0 to 30. This shows the limit cycle, which we can see more clearly if we restrict the domain to [20, 30].

```
PhasePlot[{x'[t] == x[t] - y[t] - x[t]^3, y'[t] == x[t]},
  {x[t], y[t]}, {t, 0, 30}, {x, -2, 2}, {y, -2, 2},
  InitialValues -> {0.1, 0}];
```

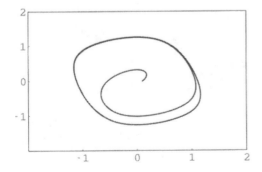

```
PhasePlot[{x'[t] == x[t] - y[t] - x[t]^3, y'[t] == x[t]},
  {x[t], y[t]}, {t, 0, 30}, {x, -2, 2}, {y, -2, 2},
  InitialValues -> {0.1, 0, {20, 30}}];
```

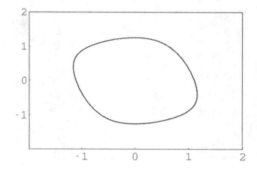

Further examples of restricting the plotting domain in this way will occur in section 4.8 on three-dimensional systems.

Here is a complicated example for which it is hard to know in advance what the proper *t*-interval is for a specific orbit. Trial and error yields that *t* going from 0 to 7.5 is sufficient to get a picture of the orbits desired.

```
PhasePlot[
  {x'[t] == x[t] - y[t]^2 Cos[y[t]], y'[t] == -y[t] + x[t] Sin[x[t]]},
  {x[t], y[t]}, {t, 0, 7.5}, {x, -10, 10}, {y, -10, 5},
  InitialValues -> {{-8.5, -10},  {-8.3, -10},
                    {-8.2, -10},  {-8,   -10},  {-7.7, -10},
                    {-7.4, -10},  {-7.3, -10},  {-7,   -10} },
  MaxSteps -> 1500];
```

```
NDSolve::mxst: Maximum number of 1500 steps reached at the
    point 4.8581.
```

```
VDS::mxstep: NDSolve ran into a MaxSteps limitation for the
    initial value {0, -8.5, -10.} and only computed the orbit
    out to t = 4.8581. If a larger domain is needed, increase
    the MaxSteps option setting.
```

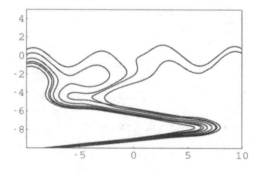

A big problem with the preceding example is that the same upper *t*-bound of 7.5 is used for all the orbits. This means not only that a lot of unnecessary computation is being done (for the parts of the orbits that are outside the window), but that error messages are likely. In fact, we are warned that the MaxSteps setting of 1500 is insufficient; yet there are no gaps in the image because the insufficiency affects only orbit-parts outside the viewing window. The reader can redo the computation with PlotRange set to All to see the big picture.

As with VisualDSolve, we can, with a lot of trial and error, find restricted plotting limits and insert them with the initial values, as follows.

```
PhasePlot[{x'[t] == x[t] - y[t]^2 Cos[y[t]],
          y'[t] == -y[t] + x[t] Sin[x[t]]},
  {x[t], y[t]}, {t, 0, 7.5}, {x, -10, 10}, {y, -10, 10},
  InitialValues -> {{-4.36, -2.47, {0, 1.3}},
      {-5.03, -2.7, {0, 1.57}},   {-6.33,   10,    {0, 7.5}},
      {-6.39,   10,   {0, 7.5}},   {-8.5,   -10,    {0, 1.3}},
      {-8.3,   -10,   {0, 1.4}},   {-8.2,   -10,    {0, 1.4}},
      {-8.0,   -10,   {0, 1.4}},   {-7.7,   -10,    {0, 4}},
      {-7.4,   -10,   {0, 4.8}},   {-7.3,   -10,    {0, 7.25}},
      {-7.0,   -10,   {0, 3.1}},   {-5.6,   -4.35, {0, 1.6}}  },
  ShowInitialValues -> True];
```

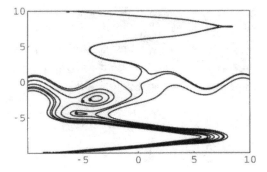

But the work involved in searching for optimal *t*-limits is tedious. We can ask PhasePlot to find these values for us by using the StayInWindow option. When that is set to True (default is False), then the orbit computations shut off automatically when the window is exited. If the window is never exited then the search uses the *t*-bounds given in the main *t*-iterator, except that it searches in both directions. Thus for the three nonexiting orbits in the following example the *t*-domain used will be from −7.5 to 7.5. When this option is turned on, the output contains the window-exit data for each initial value in a form that can be used as a setting to InitialValues.

```
PhasePlot[{x'[t] == x[t] - y[t]^2 Cos[y[t]],
          y'[t] == -y[t] + x[t] Sin[x[t]]},
  {x[t], y[t]}, {t, 0, 7.5}, {x, -10, 10}, {y, -10, 10},
  InitialValues -> {{-4.36, -2.47}, {-5.03, -2.7},
      {-6.33,   10}, {-6.39,   10}, {-8.5,   -10}, {-8.3,   -10},
      {-8.2,   -10}, {-8.0,   -10}, {-7.7,   -10}, {-7.4,   -10},
      {-7.3,   -10}, {-7.0,   -10}, {-5.6,   -4.35}},
  StayInWindow -> True]
```

```
{-Graphics-, {{0, -4.36, -2.47, {-7.5, 7.5}},
    {0, -5.03, -2.7, {-7.5, 7.5}}, {0, -6.33, 10., {0., 6.4575}},
    {0, -6.39, 10., {0., 7.31924}}, {0, -8.5, -10., {0., 1.24061}},
    {0, -8.3, -10., {0., 1.26377}}, {0, -8.2, -10., {0., 1.28853}},
    {0, -8., -10., {0., 1.38178}}, {0, -7.7, -10., {0., 3.51314}},
    {0, -7.4, -10., {0., 4.73943}}, {0, -7.3, -10., {0., 7.22243}},
    {0, -7., -10., {0., 3.08645}}, {0, -5.6, -4.35, {-7.5, 7.5}}}}}
```

If one wishes to regenerate the image, it is efficient to use the window-exit data, since that saves the solver from constantly checking to see if the window has been exited. The following code produces an image identical to the previous one.

```
PhasePlot[
    {x'[t] == x[t] - y[t]^2 Cos[y[t]], y'[t] == -y[t] + x[t] Sin[x[t]]},
    {x[t], y[t]}, {t, 0, 7.5}, {x, -10, 10}, {y, -10, 10},
    InitialValues -> %[[2]]];
```

Of course, an image might leave the viewing window only to return a short while later. The ComputeWindow option works for PhasePlot exactly as it does for VisualDSolve (discussed on page 10).

4.6   Varying a Parameter

Differential equations often involve parameters and one is very interested in how changes in the parameters affect the solutions. Here is an example of how one can create an image with different parameter values. The example, from [BCB, p. 164], is a predator-prey model with harvesting. Borrelli *et al* discuss this problem in detail and have made an interesting observation about one aspect of the model that is apparently not fully understood. Let *H* denote the harvesting constant. We will define a function that allows *H* to vary, but first lets look at a single orbit.

```
H = 0.5;
PhasePlot[{x'[t] == (-1 + y[t]) x[t] - H x[t],
          y'[t] == (1 - x[t]) y[t] - H y[t]},
  {x[t], y[t]}, {t, 0, 15}, {x, 0, 2}, {y, 0, 3.5},
  AxesLabel -> {Predator, Prey},
  Ticks -> {{0.5, 1, 1.5}, Range[3]},
  InitialValues -> {0.5, 1}];
```

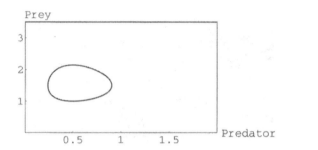

To look at more orbits, we define a function, PredPrey, of the harvesting constant, using the same initial value in each case. We set DisplayFunction to Identity to suppress the graphics at the time the images are computed. Then we map PredPrey onto a stack of harvesting constants.

```
PredPrey[H_] := PhasePlot[
  {x'[t] == (-1 + y[t]) x[t] - H x[t],
   y'[t] ==  (1 - x[t]) y[t] - H y[t]},
   {x[t], y[t]}, {t, 0, 15}, {x, 0, 2}, {y, 0, 3.5},
   DisplayFunction -> Identity, AxesLabel -> {Predator, Prey},
   PlotStyle -> {AbsoluteThickness[1]},
   InitialValues -> {0.5, 1}, ShowInitialValues -> True,
   InitialPointStyle -> AbsolutePointSize[6],
   FastPlotting -> True]

Show[Map[PredPrey, Join[Range[0, 1, 0.1], {1.5, 2, 3}]],
   DisplayFunction -> $DisplayFunction,
   Ticks -> {{0, 0.5, 1, 1.5}, {1, 2, 3}}];
```

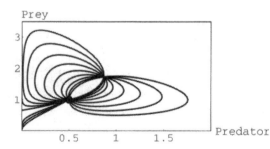

Of course, all these orbits pass through the initial point, (0.5, 1). What is surprising is that they all pass through another point, near (0.85, 1.7).

4.7 ■ Nullcline Plots

A Detailed Example

Just as with `VisualDSolve` command, we can get lots of information about the solutions without actually computing them. Direction fields are useful, but exact curves are even more useful. The `Nullclines` option shows the curves where either of the two equations is zero. More precisely, if the system is autonomous and has the form $dx/dt = f(x, y)$, $dy/dt = g(x, y)$, then the nullclines are the curves defined by $f(x, y) = 0$ or $g(x, y) = 0$. These are computed by superimposing two contour plots. Because there is no need for a t-domain, that iterator can be omitted. The `NullclineStyle` option can be used to control the graphics instructions for the x- and y-nullclines; it takes two lists. Thus the image that follows has a thick line for the x-nullcline (where dx/dt vanishes). Throughout this section we use a fairly complicated example from [BCB] which has 16 equilibrium points in the viewing window.

```
system1 = {x'[t] == x[t] - y[t]^2 Cos[y[t]],
           y'[t] == -y[t] + x[t] Sin[x[t]]};

PhasePlot[system1, {x[t], y[t]}, {x, -10, 10}, {y, -10, 10},
   Nullclines -> True,
   NullclineStyle -> {{AbsoluteThickness[2]}, {}}];
```

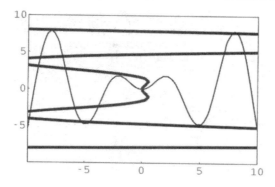

The preceding diagram shows the places where the orbits go straight up or down ($x' = 0$, given in this case by $x = y^2 \cos y$) or left or right ($y' = 0$; in this case: $y = x \sin x$). There are several noteworthy points about the geometry of these curves. First, the places where the curves intersect (more precisely, where one curve intersects another, as opposed to a self-intersection, which can arise) are the places where both x' and y' are 0. Such points are called equilibrium points and they form orbits by themselves. `PhasePlot` can compute these points exactly and show them.

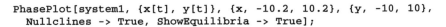

```
PhasePlot[system1, {x[t], y[t]}, {x, -10.2, 10.2}, {y, -10, 10},
    Nullclines -> True, ShowEquilibria -> True];
```

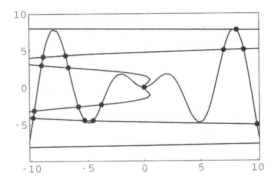

For this example there are 16 equilibria and the routine successfully finds them all. If we wish to see them explicitly, we can use the `Equilibria` function by itself, noting that it takes two expressions as its first argument (as opposed to two equations).

```
Equilibria[{x - y^2 Cos[y], -y + x Sin[x]},
    {x, -10, 10}, {y, -10, 10}]
```

```
{{-9.0854, 3.02456}, {-6.70018, 2.71369}, {0, 0}, {-3.80395, -2.33935},
    {-5.81937, -2.60337}, {-9.75079, -3.1229}, {7.75872, 7.72355},
    {8.199, 7.71582}, {-8.93902, 4.17347}, {-6.95566, 4.33284},
    {7.06874, 4.99915}, {8.81273, 5.0633}, {-9.85162, -4.07856},
    {-5.28544, -4.44109}, {-4.54663, -4.48432}, {9.96273, -5.1047}}
```

There are some options to `Equilibria`. Some of them will hardly ever be used, but for completeness we mention them. This requires a knowledge of how the equilibria-finding routine works. Briefly, to solve $f = 0$ and $g = 0$ we obtain the data set making up the $f = 0$ curve by using `ContourPlot` with `Contours` set to {0}, and then we evaluate g on that data set. Sign-changes of g correspond to points that are quite close to the solutions. We use them as seeds to *Mathematica*'s `FindRoot` command. The `SeedsOnly` option, which is primarily a debugging tool, returns the seeds only and skips the root-finding step. This might be adequate for some quick investigations, as it would save a small amount of time. The `EquilibriaPlotPoints` option has a default of 49, which means that a `PlotPoints` setting of 49 is used to generate the contour plot of $f = 0$; complicated problems might require a higher setting. And some elimination of duplicates is required since it is possible for different seeds to lead to the same root. The `EquilibriaPrecisionGoal` option controls the precision underlying the removal of duplicates.

Finally, the `ShowEigenvalues` option causes the display of a table of the equilibria with the associated eigenvalues of the system. Of course, this option can be included in `PhasePlot`.

```
Equilibria[{x - y^2 Cos[y], -y + x Sin[x]},
    {x, -10, 10}, {y, -10, 10}, ShowEigenvalues -> True];
```

Equilibria	Eigenvalues
{-9.1, 3.}	{-7.7, 7.7}
{-6.7, 2.7}	{7.15601 I, -7.15601 I}
{0, 0}	{-1, 1}
{-3.8, -2.3}	{4.99759 I, -4.99759 I}
{-5.8, -2.6}	{-6.2, 6.2}
{-9.8, -3.1}	{7.7734 I, -7.7734 I}
{7.8, 7.7}	{-10., 10.}
{8.2, 7.7}	{10.1554 I, -10.1554 I}
{-8.9, 4.2}	{8.85039 I, -8.85039 I}
{-7., 4.3}	{-9.3, 9.3}
{7.1, 5.}	{12.324 I, -12.324 I}
{8.8, 5.1}	{-14., 14.}
{-9.9, -4.1}	{9., -9.}
{-5.3, -4.4}	{5.71553 I, -5.71553 I}
{-4.5, -4.5}	{-5.6, 5.6}
{10., -5.1}	{15.8977 I, -15.8977 I}

A very important enhancement to the nullcline curves comes from considering the regions defined by the curves and their significance. A moment's thought should convince you that within each region the direction field points in one of four possible directions: northeast, southeast, northwest, or southwest. PhasePlot can color these regions in four shades of gray. To enhance the resolution we increase the setting of NullclinePlotPoints.

```
PhasePlot[system1, {x[t], y[t]}, {x, -10, 10}, {y, -10, 10},
    Nullclines -> True, NullclineShading -> True,
    ShowEquilibria -> True, NullclinePlotPoints -> 80];
```

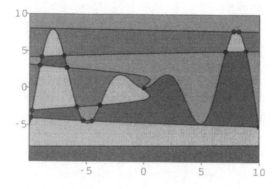

The meaning of the gray regions is clarified by the inclusion of a vector field.

```
PhasePlot[system1, {x[t], y[t]}, {x, -10, 10}, {y, -10, 10},
   Nullclines -> True, NullclineShading -> True,
   ShowEquilibria -> True, NullclinePlotPoints -> 80,
   VectorField -> True];
```

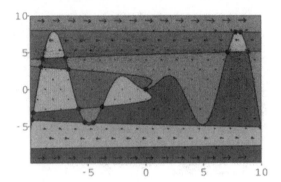

Because the vectors change length dramatically over the region in question, it is worth getting a nicer image by tweaking the FieldLength and FieldLogScale options, as discussed on page 70. Note also that, by default, tick marks are placed outside the axes in this sort of plot. If you prefer no tick marks at all, set TickLabelsOnly to True.

```
PhasePlot[system1, {x[t], y[t]}, {x, -10, 10}, {y, -10, 10},
   Nullclines -> True,  NullclineShading -> True,
   NullclinePlotPoints -> 80, ShowEquilibria -> True,
   FieldLogScale -> Automatic,
   VectorField -> True, TickLabelsOnly -> True, FieldLength -> 1.4];
```

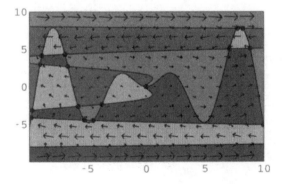

This shows clearly how the shaded regions correspond to the forward directions of motion. We can of course use colored fish, with the Hue setting causing the color to be determined by the strength of the flow. The output of the following command is in color plate 11.

```
PhasePlot[system1, {x[t], y[t]}, {x, -10, 10}, {y, -10, 10},
   Nullclines -> True, NullclineShading -> True,
   ShowEquilibria -> True, NullclinePlotPoints -> 80,
   FlowField -> True, FieldLogScale -> Automatic,
   FieldLength -> 1.2, FieldThickness -> 0.7, FieldColor -> Hue];
```

Note how much information we are getting about the situation without yet having plotted a single orbit! When we do look at the actual orbits, in color, superimposed on the gray background, not only is the resulting image visually pleasing, it is also much more informative than an image of the orbits alone, or with a direction field. In the image that follows we use the `ParametricPlotFunction` option to color the orbits according to increasing *t*-values. We also specify the *t*-limits, using information gleaned from the work on page 78 (the color image is in color plate 12).

```
PhasePlot[system1, {x[t], y[t]}, {t, 0, 7.5},
   {x, -10, 10}, {y, -10, 10},
   Nullclines -> True, NullclineShading -> True,
   ShowEquilibria -> True, EquilibriumPointStyle -> Red,
   NullclinePlotPoints -> 80,
   InitialValues -> { {-4.36, -2.47, {0, 1.30}},
                      {-5.03, -2.7,  {0, 1.57}},
                      {-6.33, 10,    {0, 6.46}},
                      {-6.39, 10,    {0, 7.32}},
                      {-8.5, -10,    {0, 1.25}},
                      {-8.3, -10,    {0, 1.27}},
                      {-8.2, -10,    {0, 1.29}},
                      {-8., -10,     {0, 1.39}},
                      {-7.7, -10,    {0, 3.52}},
                      {-7.4, -10,    {0, 4.74}},
                      {-7.3, -10,    {0, 7.23}},
                      {-7., -10,     {0, 3.09}},
                      {-5.6, -4.35,  {0, 1.60}} },
   ParametricPlotFunction -> ColorParametricPlot,
   PlotPoints -> 400,
   PlotStyle -> Thickness[0.0017]];
```

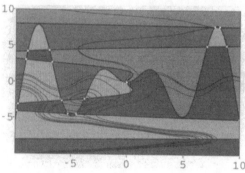

The algorithm we use to get the gray regions is very complicated! It is explained in [SW]. In brief, we exploit the fact that *Mathematica* allows us to access the data in a contour plot. So once the contour plots of the nullclines are in hand it is, in theory, possible to obtain the boundaries of the various regions from that data. Suffice it to say that, in practice, this approach is incredibly complicated. There is a simpler approach to the gray regions that is worth mentioning. Its drawbacks are that it is slow and has resolution problems near the nullclines. Simply form a density plot of 3sign f + sign g. Because this weighted sum of signs takes on only integer values between -4 and 4, inclusive, and distinct values correspond to distinct directions, this does the job. But the complicated method we actually used in `NullclinePlot` produces better images and is fairly fast. In fact, we have implemented an option that forces the sum-of-signs method to be used; some examples of its use are given in the subsection that follows.

Miscellaneous Examples

We conclude this section with some miscellaneous examples that illustrate, among other things, that the nullcline shading program can handle a variety of behaviors. The following example is unusual in that one of the nullclines crosses itself; the crossing point is not an equilibrium point. This example illustrates an important feature of `StayInWindow`, for when that is set the orbits are drawn in both directions, and the main *t*-iterator is irrelevant. Thus the user need not know which direction is better for the particular window and starting values.

```
system2 = {x'[t] == x[t]^3 - 4 x[t] y[t] + 2 y[t]^2 + y[t]^3,
           y'[t] == -1 + x[t]^2 + y[t]};

PhasePlot[system2, {x[t], y[t]},
   {t, 0, 10}, {x, -2, 2}, {y, -2, 2},
   Nullclines -> True, NullclineShading -> True,
   ShowEquilibria -> True];
```

It sometimes takes some effort to find a good selection of initial values. Since there are three equilibria, we can take three sets of initial values, each set surrounding an equilibrium point. To get the three sets, we first get the exact equilibrium points. Then we map a pure function that uses `Table` to get 12 points in a circle of radius 0.2 around each one.

```
eqs = Equilibria[{x^3 - 4 x y + 2 y^2 + y^3, -1 + x^2 + y},
                {x, -2, 2}, {y, -2, 2}];

threeCircles = N[Flatten[Map[Table[# + 0.2 {Cos[t], Sin[t]},
    {t, 0, 2 Pi - Pi/6, Pi/6}] &, eqs], 1], 2];
```

Now we can look at the 36 orbits; well, 37 since we include (0, 0) to see what happens at the self-intersection point. The output is in color plate 13.

```
PhasePlot[system2, {x[t], y[t]}, {t, 0, 10},
    {x, -2, 2}, {y, -2, 2},
    Nullclines -> True, NullclineShading -> True,
    NullclinePlotPoints -> 80, Rainbow -> True,
    FastPlotting -> True, ShowEquilibria -> True,
    ShowEigenvalues -> True, StayInWindow -> True,
    EquilibriumPointStyle -> {AbsolutePointSize[6], White},
    InitialValues -> Append[threeCircles, {0, 0}]];
```

Equilibria	Eigenvalues
{-0.9, 0.18}	{1.35709 + 2.81311 I, 1.35709 - 2.81311 I}
{0.54, 0.71}	{-2.6, 1.7}
{0.88, 0.23}	{1.20972 + 2.0715 I, 1.20972 - 2.0715 I}

Here's an example that is included only because it produces a pretty image (see color plate 14).

```
PhasePlot[{x'[t] == Exp[-x[t]^4 - y[t]^4],
            y'[t] == (x[t]^2 + y[t]^2 - 1) Exp[-x[t]^4 - y[t]^4]},
    {x[t], y[t]}, {x, -2, 2}, {y, -2, 2},
    FlowField -> True, NullclineShading -> True,
    FieldColor -> Hue, AspectRatio -> Automatic];
```

The following example shows that the nullcline shading algorithm can handle a closed contour.

```
PhasePlot[{x'[t] == x[t]^3 + y[t]^3 - 4 x[t] Sin[y[t]^2] + 2 y[t]^2,
            y'[t] == y[t] + Cos[x[t]^2]^2 - 1},
    {x[t], y[t]}, {x, -2, 2}, {y, -2, 2},
    Nullclines -> True, NullclineShading -> True,
    NullclinePlotPoints -> 40, ShowEquilibria -> True];
```

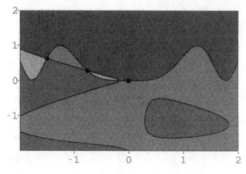

Well, actually this image is not so good. There is no nullcline-crossing shown at the central equilibrium point, and it looks like the crossing just left of the central equilibrium has been missed. The plausible interpretation is that the crossing just left of the origin should really be at the origin. Of course, examining the equations shows that the origin is indeed an equilibrium. To investigate further we zoom in and improve the resolution. This adds more evidence for the existence of a single equilibrium in the vicinity of the origin, and shows the cusp-like nature of the nullcline.

```
PhasePlot[{x'[t] == x[t]^3 + y[t]^3 - 4 x[t] Sin[y[t]^2] + 2 y[t]^2,
          y'[t] == y[t] + Cos[x[t]^2]^2 - 1},
   {x[t], y[t]}, {x, -0.3, 0.3}, {y, -0.3, 0.3},
   Nullclines -> True, NullclineShading -> True,
   NullclinePlotPoints -> 100,
   ShowEquilibria -> True;
```

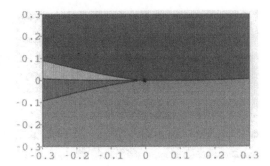

Now we look at some orbits. Note that this requires including the *t*-iterator, which is missing from the two preceding inputs. We increase the nullcline resolution to 100, which almost eliminates the small discrepancy at the origin. The output is in color plate 15.

```
PhasePlot[{x'[t] == x[t]^3 + y[t]^3 - 4 x[t] Sin[y[t]^2] + 2 y[t]^2,
          y'[t] == y[t] + Cos[x[t]^2]^2 - 1},
   {x[t], y[t]}, {t, -10, 10}, {x, -2, 2}, {y, -2, 2},
   Nullclines -> True, NullclineShading -> True,
   NullclinePlotPoints -> 100,
   ShowEquilibria -> True,
   EquilibriumPointStyle -> White,
   InitialValues -> Join[Table[{0, x, -0.4}, {x, -2, 2, 0.2}],
                         Table[{0, x,  0.4}, {x, -2, 2, 0.2}]],
   StayInWindow -> True,
   Rainbow -> True, FastPlotting -> True];
```

The next example shows that these routines are able to handle functions such as absolute value. Because Abs cannot be differentiated Equilibria calls an alternative root-finding procedure to find the equilibrium points.

```
PhasePlot[{x'[t] == 0.7 - Abs[Sin[x[t] y[t]]],
        y'[t] == x[t]^2 + y[t]^2 - 3},
    {x[t], y[t]}, {x, -3, 3}, {y, -3, 3},
    Nullclines -> True, ShowEquilibria -> True,
    NullclineShading -> True];
```

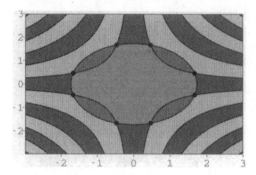

Here's an example that has a whole line of equilibria. In this case, the shading program detects the fact that Equilibria has problems (the error messages) and automatically switches to the weighted sum of signs method, eventually producing the correct regions.

```
PhasePlot[{x'[t] == x[t],  y'[t] == x[t] y[t]},
    {x[t], y[t]}, {x, -1, 1}, {y, -1, 1},
    NullclineShading -> True, Nullclines -> True];
```

```
Solve::svars: Warning: Equations may not give solutions for
all "solve" variables.
```

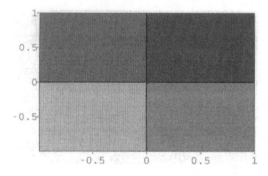

Finally, here's a troublesome example using a discontinuous function, UnitStep (which is 1 for positive arguments and 0 elsewhere). Although we could define the function with an If statement, we instead load it in via the DiracDelta package.

```
Needs["Calculus`DiracDelta`"];
```

```
Equilibria[{UnitStep[1 - x^2 - y^2] y, UnitStep[1 - x^2 - y^2] (-x)},
    {x, -2, 2}, {y, -2, 2}]
```

```
{{0, 0}}
```

No error messages were produced, but the two-dimensional region of equilibria (exterior of unit disk) has not been found. Nevertheless, we can get a proper, if poorly resolved, image by forcing the SumOfSigns algorithm to be used.

```
PhasePlot[{x'[t] == y[t] UnitStep[1 - x[t]^2 - y[t]^2],
        y'[t] == -x[t] UnitStep[1 - x[t]^2 - y[t]^2]},
    {x[t], y[t]}, {x, -2, 2}, {y, -2, 2},
    Nullclines -> True, ShowEquilibria -> True,
    NullclineShading -> True, NullclinePlotPoints -> 60,
    NullclineShadingMethod -> SumOfSigns, AspectRatio -> Automatic];
```

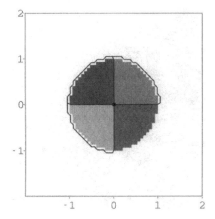

This is by no means a perfect image—the fourth-quadrant nullcline is missing—but the gray and white regions are roughly correct.

Finally, here is a very complex example that forms the cover of the Spring 1995, issue of the newsletter of the Consortium for Ordinary Differential Equations Experiments (CODEE). Compare the information in the shaded region plot to what can be gleaned from the orbits alone. The shaded plot takes some time to generate. But it does show that the shaded nullcline algorithm can handle a very complicated situation. Note the latent image of a bear's head that shows up at the bottom of the shaded nullcline plot!

```
system3 = {x'[t] == -Cos[y[t]] + 2 y[t] * Cos[y[t]^2] * Cos[2 x[t]],
        y'[t] == -Sin[x[t]] + 2 Sin[y[t]^2] * Sin[2 x[t]]};

PhasePlot[system3, {x[t], y[t]},
    {t, -10, 10}, {x, -3.45, 4}, {y, -1.8, 4.2},
    MaxSteps -> 1500, StayInWindow -> True,
    InitialValues -> Join[{{0, Pi, 1.7}, {0, Pi, 2.4}},
        Table[{0, 0, i}, {i, -0.5, 3, 0.5}]],
    FastPlotting -> True, AspectRatio -> 1.6];
```

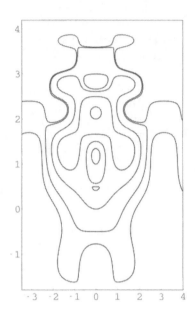

```
PhasePlot[system3, {x[t], y[t]},
   {t, -10, 10}, {x, -3.45, 4}, {y, -1.8, 4.2},
   MaxSteps -> 1500, AspectRatio -> 1.6,
   InitialValues -> Join[{{0, Pi, 1.7}, {0, Pi, 2.4}},
                         Table[{0, 0, i}, {i, -0.5, 3, 0.5}]]],
   Nullclines -> True, NullclinePlotPoints -> 120,
   ShowEquilibria -> True, FastPlotting -> True,
   StayInWindow -> True, NullclineShading -> True];
```

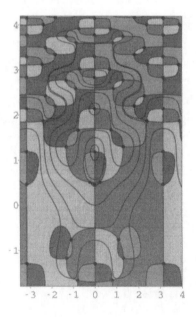

4.8 ■ The Real Orbit Graph in Three Dimensions

The package contains a function called `ProjectionPlot3D` that takes a parametrically defined curve in a plane and produces a three-dimensional image of the graph, together with the three projections onto the coordinate planes. The function is really meant for internal use, as an option setting to `ParametricPlotFunction` within `PhasePlot`. Thus we must set the `DisplayFunction` to override the `Identity` setting, which suppresses the display. Here is an example that shows a 4-leafed rose (color output in color plate 16).

```
ProjectionPlot3D[Sin[4 t] {Cos[t], Sin[t]},
   {t, -2 Pi, 0}, {-1.5, 1.5}, {-1.5, 1.5},
   PlotPoints -> 60, PlaneResolution -> 4,
   MainThickness -> 0.008, BoxRatios -> {1, 1, 1},
   AxesLabel -> {x, y, t}, DisplayFunction -> $DisplayFunction];
```

There are some options that allow customization. One can turn the color off, change the thickness of the curves, and change the resolution of the planes.

```
Options[ProjectionPlot3D]

{MainThickness -> 0.008, MinorThickness -> 0.001, PlotPoints -> 40,
   PlaneResolution -> 7, Color -> True}
```

Here is a very simple example. The solution is a circle, whose graph in 3-space is a helix (color output in color plate 17).

```
PhasePlot[{x'[t] == -y[t], y'[t] == x[t]}, {x[t], y[t]},
   {t, 0, 10 Pi}, {x, -1.2, 1.2}, {y, -1.2, 1.2},
   InitialValues -> {1, 0}, BoxRatios -> {1, 1, 2},
   MainThickness -> 0.01, PlotPoints -> 200,
   ParametricPlotFunction -> ProjectionPlot3D];
```

By using a Do-loop with different `ViewPoint` settings one can, time and memory permitting, generate an animation of a three-dimensional object moving in space. For more details on how to do this, including an explanation of the coordinate system in which the viewpoints are defined, see [Wag, chap. 3] or [SB].

4.9 ■ Systems of Three or More Equations

We can look at systems of three or more equations, and either look at two of the variables in the plane, or three of them in three-space. Our first example is a Rössler system. By specifying two of the x, y, and z iterators, we get a view in the specified plane. Note that in this situation we must give the list of dependent variables as an argument following the equation.

```
Rossler = {x'[t] == -y[t] - z[t],
           y'[t] == x[t] + 0.41 y[t],
           z'[t] == 2 - (4 - x[t]) z[t]};

PhasePlot[Rossler, {x[t], y[t], z[t]}, {t, 0, 50},
  {x, -4, 6}, {y, -6, 4}, InitialValues -> {0, 2, 0},
  MaxSteps -> 1000, FastPlotting -> True, AxesLabel -> {x, y}];
```

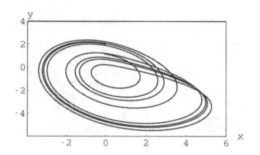

```
PhasePlot[Rossler, {x[t], y[t], z[t]},
  {t, 0, 50}, {x, -4, 6}, {z, 0, 8}, InitialValues -> {0, 2, 0},
  MaxSteps -> 1000, FastPlotting -> True, AxesLabel -> {x, z}];
```

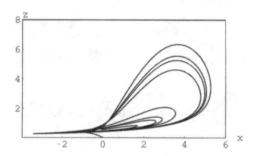

```
PhasePlot[Rossler, {x[t], y[t], z[t]},
  {t, 0, 50}, {y, -6, 4}, {z, 0, 8}, InitialValues -> {0, 2, 0},
  FastPlotting -> True, MaxSteps -> 1000, AxesLabel -> {y, z}];
```

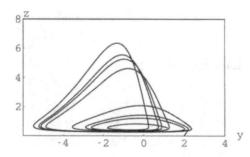

And by specifying all three iterators, we get the orbit in *xyz*-space. FastPlotting works here as well; it causes the image to be generated by simply connecting the points found by the numerical differential equation solver (as opposed to calling Plot3D).

```
PhasePlot[Rossler, {x[t], y[t], z[t]},
    {t, 0, 50}, {x, -4, 6}, {y, -6, 4}, {z, 0, 8},
    InitialValues -> {0, 2, 0}, FastPlotting -> True,
    MaxSteps -> 1000];
```

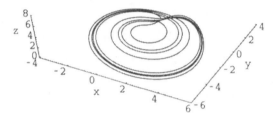

We can check out the longer term by solving the equation out to *t* = 500 but showing the plot only for the interval from 450 to 500, which we add as an extra value to the initial data.

```
PhasePlot[Rossler, {x[t], y[t], z[t]},
    {t, 0, 500}, {x, -4, 6}, {y, -6, 4}, {z, 0, 8},
    InitialValues -> {0, 2, 0, {450, 500}},
    FastPlotting -> True, MaxSteps -> 1000];
```

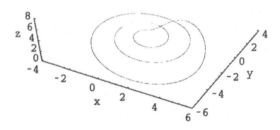

```
Show[%, ViewPoint -> {0.2, -2.6, -0.07}];
```

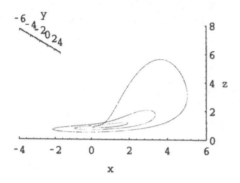

The following example from [BCB, p. 329] is a model of changing voltages and current in an electric circuit. It shows the advantage of color and viewpoint selection. For example, if we use black and white with the default viewpoint, it is fairly hard to see what is going on. Warning: This example consumes a lot of memory; use a lower *t*-bound if this is a concern.

```
Clear[f];

f[a_, w_, b_] := -a w + (a + b)/2 (Abs[w + 1] - Abs[w - 1])

PhasePlot[{x'[t] == -f[7/100, y[t] - x[t], 1/10] / 10,
            y'[t] == -f[7/100, y[t] - x[t], 1/10] - z[t],
            z'[t] == y[t]},
   {x[t], y[t], z[t]}, {t, 0, 350},
   {x, 1.0, 1.44}, {y, -3, 3}, {z, -2.5, 2.5},
   InitialValues -> {1, 1, 1},
   FastPlotting -> True, SaveSolution -> True,
   PlotPoints -> 900, Boxed -> True, MaxSteps -> 5000];
```

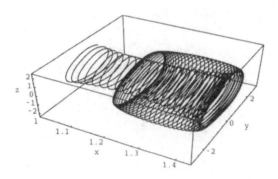

When we use color to show increasing *t*-values and change the viewpoint, the mushroom nature of the orbit is much clearer. But this approach requires a lot of memory! Some straightforward programming with Graphics3D adds gray shading to the box (output is in color plate 18).

```
orbit = PhasePlot[{x'[t] == -f[7/100, y[t] - x[t], 1/10] /10,
                   y'[t] == -f[7/100, y[t] - x[t], 1/10] - z[t],
                   z'[t] == y[t]},
  {x[t], y[t], z[t]}, {t, 0, 350},
  {x, 1.0, 1.44}, {y, -3, 3}, {z, -2.5, 2.5},
  InitialValues -> {1, 1, 1}, PlotPoints -> 900,
  Boxed -> True, PlotStyle -> AbsoluteThickness[1.5],
  ParametricPlotFunction -> ColorParametricPlot,
  ViewPoint -> {-0.25, -1.2, 0.25}, DisplayFunction -> Identity,
  MaxSteps -> 5000, AxesEdge -> {{-1, -1}, {-1, 1}, {-1, -1}}];

Show[orbit, Graphics3D[{GrayLevel[0.8], Map[Polygon, {
  {{1, -3, -2.5}, {1, -3, 2.5}, {1, 3, 2.5}, {1, 3, -2.5}},
  {{1.44, -3, -2.5}, {1.44, -3, 2.5}, {1.44, 3, 2.5}, {1.44, 3, -2.5}},
  {{1.44, -3, -2.5}, {1.44, 3, -2.5}, {-1.44, 3, -2.5},
   {-1.44, -3, -2.5}}, {{1, 3, -2.5}, {1.44, 3, -2.5}, {1.44, 3, 2.5},
   {1, 3, 2.5}}}]}]],
  DisplayFunction -> $DisplayFunction, Lighting -> False,
  PlotRange -> {{0.99999, 1.44001}, {-3.01, 3.01}, {-2.5, 2.5}}];
```

4.10 ■ Poincaré Sections

PoincareSection constructs Poincaré sections of first order systems of two equations or a single second-order equation. It handles a single initial value, which must be passed as an argument, not an option. The default situation shows 101 points.

```
PoincareSection[{x'[t] == y[t], y'[t] == -x[t] - y[t]/10},
  {t, 0, 5 Pi}, {x, -1.4, 1.4}, {y, -1.4, 1.4}, {0, 1}];
```

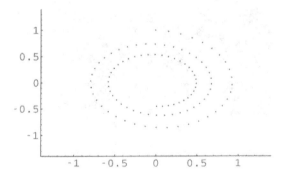

The TimeInterval option controls the interval at which points are sampled. We also bumped up the MaxSteps option to 1000 since Poincaré sections often use large t-intervals requiring a large number of steps to solve.

```
PoincareSection[{x'[t] == y[t], y'[t] == -x[t] - y[t]/10},
   {t, 0, 100 Pi}, {x, -1.4, 1.4}, {y, -1.4, 1.4}, {0, 1},
   TimeInterval -> Pi/2];
```

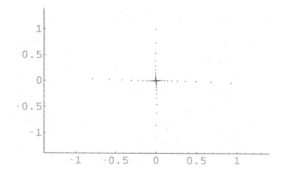

Here is an example that is based on a second-order Duffing equation, discussed in much more detail in Chapter 15. If the Rainbow option is turned on then the points are drawn using hues that change with increasing time (from Hue[0] (red) to Hue[0.5] (cyan)). In the example that follows the later hues cover up the early ones. Section 5.1 contains some images that show the relationship between this Poincaré section and the Duffing orbit. The output of the following command is in color plate 19.

```
PoincareSection[
   {x'[t] == u[t], u'[t] == Cos[2 t]/4 - x[t] - x[t]^3},
   {t, 0, 100 Pi}, {x, 2, 2.15},
   {u, -1.15, 1.15}, {2.03284, 0},
   TimeInterval -> Pi, AspectRatio -> 0.4,
   MaxSteps -> 10000, Rainbow -> True,
   PointStyle -> AbsolutePointSize[5],
   WindowShade -> GrayLevel[0.5]];
```

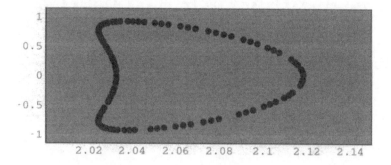

There are times when one might wish to have the points of the Poincaré section saved for further use. The SavePoints option causes the set of points to be returned.

```
PoincareSection[
  {x'[t] == u[t], u'[t] == Cos[2 t]/4 - x[t] - x[t]^3},
  {t, 0, 10 Pi}, {x, 2, 2.15},
  {u, -1.15, 1.15}, {2.03284, 0},
  TimeInterval -> Pi, DisplayFunction -> Identity,
  SavePoints -> True]
```

{-Graphics-, {{2.03284, 0.}, {2.02779, 0.507409},
    {2.02542, 0.846981}, {2.04566, 0.913584},
    {2.08562, 0.676658}, {2.116, 0.195851},
    {2.10983, -0.359924}, {2.0728, -0.77944},
    {2.03665, -0.92531}, {2.02446, -0.768801},
    {2.03009, -0.363496}}}

Chapter 5

SecondOrderPlot

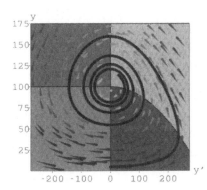 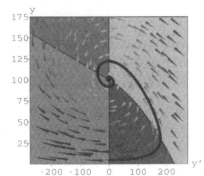

Overview

This chapter discusses the use of SecondOrderPlot to study the solutions to one or more second-order equations. This is really the culmination of the whole package since this function makes use of many of the other comprehensive functions such as PhasePlot and SystemSolutionPlot, and also because this class of equations encompasses so many fascinating examples.

FUNCTIONS DISCUSSED: ToSystem, PhasePlot, PoincareSection, SecondOrderPlot.

5.1 ■ A Single Second-Order Equation

SecondOrderPlot accepts a second-order equation in the form x″[t] == x[t] + t, or x″[t] - x[t] == t. One important distinction is that the second argument must consist of the dependent function (or list of functions in the case of a system). The simplest usage is as follows. No plotting domain is given and the default is to provide graphs of x and x' in a single image.

```
SecondOrderPlot[x″[t] == -x[t], x[t], {t, 0, 2 Pi},
    InitialValues -> {0, 1}];
```

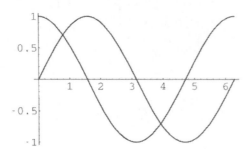

Note that the initial values must always appear in the order (x, x'), or (x, y, ..., x', y', ...) in the case of a system, and the *t*-value at which the initial values are set is the first *t*-value in the *t*-iterator. This is all a little less flexible than the initial value possibilities for PhasePlot, but the user can always transform the equation(s) to a system and use PhasePlot (an example where this is useful occurs in the helium balloon section that follows).

If separate images are desired, it can be accomplished with the PlotVariables option, which is passed through to SystemSolutionPlot (we can use PlotLabels to identify the graphs).

```
SecondOrderPlot[x"[t] == -x[t], x[t], {t, 0, 2 Pi},
    InitialValues -> {0, 1}, PlotVariables -> {{x}, {x'}},
    PlotLabels -> {"x vs. t", "x' vs. t"}];
```

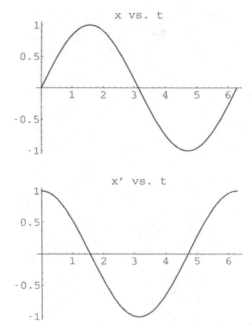

If only one graph is wanted, then that is obtained by feeding only a single plotting variable as an argument; it can be either x or x'. In these cases the plotting range would be controlled by the `PlotRange` option (unlike `VisualDSolve`, which used an iterator, {x, a, b}, to control the window height).

```
SecondOrderPlot[x"[t] == -x[t], x[t], {t, 0, 2 Pi},
   PlotVariables -> {{x'}},
   InitialValues -> {0, 1}, AxesLabel -> {Time, Speed}];
```

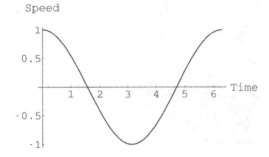

Note that `SecondOrderPlot` can interpret, say, x' properly in its role as, for example, a plotting variable. When the plotting variables are given in the form of iterators, as opposed to just variables, then they form the basis of a phase plot. Often for a second-order equation one will want to see *x* vs. *x'*. The initial values may be set to any of the forms suitable for `PhasePlot`.

```
SecondOrderPlot[x"[t] == -x[t], x[t],
   {t, 0, 2 Pi}, {x, -1.5, 1.5}, {x', -1.5, 1.5},
   InitialValues -> {{0, 1}, {0, 0.5, {0, Pi}}},
   AspectRatio -> Automatic, AxesLabel -> {x, x'}];
```

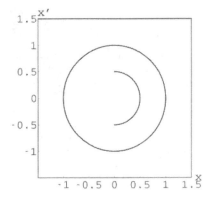

Here are some views of the phase space for a standard pendulum. We begin with the simple case where there is no friction. The undamped pendulum is a Hamiltonian system (see Chapter 10), and that allows us to figure out some initial values on the orbits that separate the closed orbits from the open ones. Thus the image includes orbits forming the separatrix.

```
SecondOrderPlot[x"[t] == -8 Sin[x[t]], x[t],
  {t, 0, 4.2}, {x, -10, 10}, {x', -10, 10},
  Nullclines -> True,
  NullclineShading -> True,
  StayInWindow -> True,
  PlotStyle -> AbsoluteThickness[0.7],
  ShowEquilibria -> True,
  EquilibriumPointStyle -> {AbsolutePointSize[5], White},
  InitialValues -> Join[
     { {     0, -4 Sqrt[2]}, {     0, 4 Sqrt[2]},
       { 2 Pi, -4 Sqrt[2]}, { 2 Pi, 4 Sqrt[2]},
       {-2 Pi, -4 Sqrt[2]}, {-2 Pi, 4 Sqrt[2]}},
     Table[{Pi, y}, {y,  -6.5, -1.5}],
     Table[{Pi, y}, {y,   1.5,  6.5}],
     Table[{ x, 0}, {x, -2 Pi,  -Pi, Pi/4}],
     Table[{ x, 0}, {x,    0,    Pi, Pi/4}],
     Table[{ x, 0}, {x,  2 Pi, 3 Pi, Pi/4}]]];
```

The next example shows the effect of adding friction. Now all the orbits eventually settle into one of the equilibria on the *x*-axis. Note how useful the shaded nullcline view is: the regions allow us to see at a glance where the equilibria are and the general shape of the orbits. This image also gives a sense of the shape of the basin of attraction of a single equilibrium by looking at the band of orbits that end up there.

```
SecondOrderPlot[x"[t] == - x'[t] - 10 Sin[x[t]], x[t], {t, 0, 15},
  {x, -16, 16}, {x', -16, 16},
  Nullclines -> True, NullclineShading -> True,
  NullclinePlotPoints -> 60, FastPlotting -> True,
  InitialValues -> Join[Table[{x,  15}, {x, -14, -7, 0.8}],
                        Table[{x, -15}, {x, 7.6, 14, 0.8}]]];
```

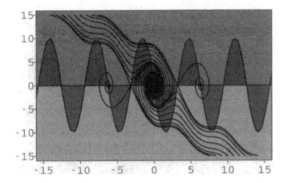

Another tool that can be used with a single second order differential equation is
`PoincareSection`. Here is an example that comes from the Duffing equation, which
is studied in much more detail in Chapter 15.

```
pSec = PoincareSection[x"[t] == Cos[2 t]/4 - x[t] - x[t]^3,
  {t, 0, 50 Pi}, {x, 2, 2.15}, {x', -1.15, 1.15}, {2.03284, 0},
  TimeInterval -> Pi, MaxSteps -> 10000,
  Ticks -> {{2.02, 2.06, 2.1, 2.14}, Range[-1, 1, 0.5]},
  PointStyle -> AbsolutePointSize[4]];
```

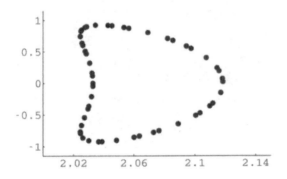

We can examine the orbit out to 10π, though the view we get gives no indication
of where the heart-like shape lies.

```
orbit = SecondOrderPlot[x"[t] == Cos[2 t]/4 - x[t] - x[t]^3,
  x[t], {t, 0, 10 Pi}, {x, -2.2, 2.2}, {x', -4, 4},
  InitialValues -> {2.03284, 0},
  FastPlotting -> True,
  AxesLabel -> {x, x'}, MaxSteps -> 1000];
```

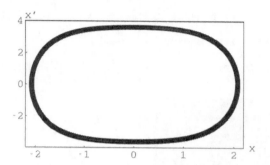

Combining the two images shows approximately where the points of the Poincaré section lie.

```
Show[orbit, pSec, PlotRange -> All];
```

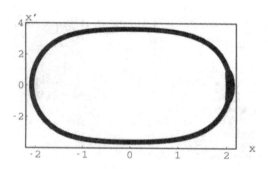

A closer look is called for.

```
Show[pSec, orbit, Axes -> Automatic, AxesOrigin -> {1.9, 0},
    Ticks -> {{2, 2.15}, Range[-2, 2]},
    PlotRange -> {{1.9, 2.15}, {-2, 2}}];
```

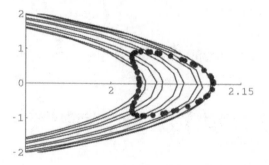

5.2 ■ The Ups and Downs of a Helium Balloon

Helium balloons provide an interesting example because, under certain conditions, they will oscillate up and down to an equilibrium. Imagine a helium-filled balloon attached to a long rope with linear density ρ. Newton's second law of motion—the time derivative of momentum is equal to the sum of the forces—yields a second-order differential equation. We denote by g the gravitational constant (980) in cm/sec^2. We use y for the height of the balloon off the ground and v for the velocity of the balloon.

There are four forces acting on the balloon. Let F_w denote the weight of the system, which is the variable weight of the part of the rope which is off the ground ($\rho \cdot y$) plus the weight of the balloon (W_b = 3 grams). F_h is the constant force from the helium in the balloon. Roughly, the balloon displaces 13 liters. Using the specific gravity of air at room temperature (1.161 grams/liter) and of helium (0.16 grams/liter), we can approximate the upward force due to the helium as F_h = g·(mass air displaced− mass helium) = $(13 \cdot 1.161 - 13 \cdot 0.16)g = 13g$ grams · cm/sec^2. F_r is air resistance on the balloon, which is a function of the velocity of the balloon, and which, for a start, we will ignore.

The fourth force is realized from the fact that the rope loses momentum as it hits the ground. Thus the ground must exert an upward force, F_g, on the rope equal to the rate at which the rope loses momentum (mv, or $\rho \cdot y \cdot v$) hitting the ground. If we consider the case of just a falling rope we get the following equation: $d(\rho \cdot y \cdot v)/dt = F_g - \rho \cdot y \cdot g$, where $-\rho \cdot y \cdot g$ is the force of gravity. Expanding the left side and using $dv/dt = -g$ (acceleration is constant for a free-falling rope) yields: $\rho \cdot v^2 - \rho \cdot y \cdot g = F_g - \rho \cdot y \cdot g$. Thus $F_g = \rho \cdot v^2$ while the rope is falling and equals 0 otherwise.

Now, still ignoring air resistance, we first define some forces and work out the form of the differential equation.

```
g = 980; rho = 0.1; Wb = 3;
Fw = -g * (Wb + rho y[t]);
Fh = 13 g;
Fg = If[y'[t] < 0, rho y'[t]^2, 0];
```

We can now set up the differential equation using Newton's second law and the product of mass and velocity, `(Wb + rho y[t]) y'[t]`, to get the left-hand side (change in momentum). We are here using *Mathematica* to differentiate the abstract expression representing momentum with respect to t.

```
balloonEqn = (D[(Wb + rho y[t]) y'[t], t] == Fh + Fw + Fg + Fr)
```

$$0.1\ y'[t]^2 + (3 + 0.1\ y[t])\ y''[t] ==$$
$$12740 + Fr + If[y'[t] < 0, rho\ y'[t]^2, 0] - 980\ (3 + 0.1\ y[t])$$

```
Fr = 0;
SecondOrderPlot[balloonEqn, y[t],
   {t, 0, 6}, {y, 0, 175}, {y', -275, 275},
   InitialValues -> {5, 0}, AspectRatio -> 1,
   FastPlotting -> True, ShowInitialValues -> True,
   AxesLabel -> {y, y'}];
```

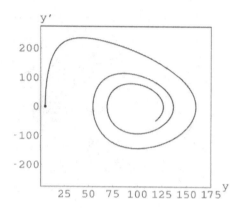

We would like to rotate this so that the vertical axis represents the height of the balloon. However, SecondOrderPlot is finicky in that it wants its initial values to have the function values first, followed by the derivative values; and the iterators have to respect this order as well. So to do what we want we resort to transforming the second-order equation to a system and using PhasePlot. This is not too onerous because we can use ToSystem (see section 2.4) to do the transformation for us. We clear Fr so we can see its effect on the equation. And we use the AuxiliaryVariables option to force v to be the auxiliary variable that represents y'.

```
Clear[Fr];
balloonSystem =
   ToSystem[balloonEqn, y[t], t, AuxiliaryVariables -> v]
```

$$\{y'[t] == v[t], \ v'[t] == -(\frac{-9800. \ - \ Fr \ - \ If[v[t] < 0, \ rho \ v[t]^2, \ 0] \ + \ 0.1 \ v[t]^2 \ + \ 98. \ y[t]}{3. \ + \ 0.1 \ y[t]})\}$$

Now we can feed balloonSystem to PhasePlot with v, the velocity, coming ahead of y at all steps, including the initial values. We start the balloon 5 cm above the ground; this could be 0, but the image is more easily interpreted if the initial condition is (0, 5) as opposed to (0, 0).

```
Fr = 0;
BalloonSansAirResistance = PhasePlot[balloonSystem,
   {v[t], y[t]}, {t, 0, 10}, {v, -275, 275}, {y, 0, 175},
   InitialValues -> {0, 5}, AspectRatio -> 1,
   FastPlotting -> True, MaxSteps -> 1000, Nullclines -> True,
   NullclineShading -> True,
   FlowField -> True, FieldColor -> GrayLevel,
   AxesLabel -> {y', y},
   FieldLogScale -> Automatic, PlotStyle -> AbsoluteThickness[2]];
```

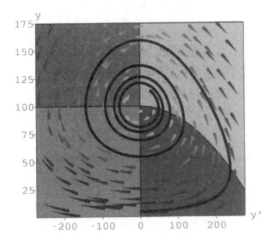

We can now add air resistance on the model by specifying F_r to be an appropriate function of v. A good model for air resistance requires experimentation to calculate a function of velocity and makes for a good student project. Here we simply choose a linear function that roughly matches what happens when a balloon attached to a heavy string is set in motion. We omit the resulting single image, showing instead a side-by-side graphic, which gives a striking display of how resistance is reflected in the three components of the phase plots.

```
Fr = - 20 v[t];
BalloonWithAirResistance = PhasePlot[balloonSystem,
   {v[t], y[t]}, {t, 0, 10}, {v, -275, 275}, {y, 0, 175},
   InitialValues -> {0, 5}, AspectRatio -> 1,
   FastPlotting -> True, MaxSteps -> 1000, Nullclines -> True,
   NullclineShading -> True,
   FlowField -> True, FieldColor -> GrayLevel,
   AxesLabel -> {y', y},
   FieldLogScale -> Automatic, PlotStyle -> AbsoluteThickness[2]];
```

```
Show[GraphicsArray[
   {BalloonSansAirResistance, BalloonWithAirResistance}]];
```

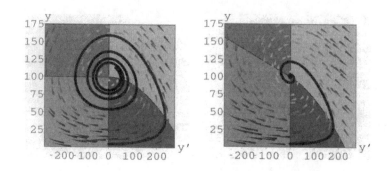

5.3 ▮ Second-Order Systems

We illustrate second-order systems by looking at Newton's laws applied to planetary motion. We can consider a planar cross-section, with an eccentricity of 0.5. Since $\mathbf{X}'' = -\mathbf{X}/|\mathbf{X}|^{3/2}$ governs the motion, where $\mathbf{X}$ represents the position vector (x, y), this is easily handled by SecondOrderPlot. Since we want to see an x-y plot, we use those two variables in the iterators. And, to illustrate Kepler's law (a planet sweeps out equal area in equal time), we color the orbit in terms of its speed. The Epilog option is used to add the sun as a large point. Note that the initial values (four in this case) are understood to be in the order x, y, x', y'. The color output is in color plate 20.

```
e = 0.5;
SecondOrderPlot[{x"[t] == -x[t] / (x[t]^2 + y[t]^2)^(3/2),
                 y"[t] == -y[t] / (x[t]^2 + y[t]^2)^(3/2)},
   {x[t], y[t]}, {t, 0, 2 Pi}, {x, -2, 1}, {y, -1.4, 1.4},
   InitialValues -> {1 - e, 0, 0, Sqrt[(1 + e)/(1 - e)]},
   ParametricPlotFunction -> ColorParametricPlot,
   PlotStyle -> AbsoluteThickness[2], Speed -> True,
   WindowShade -> GrayLevel[0.5],
   Epilog -> {Yellow, AbsolutePointSize[10], Point[{0, 0}]}];
```

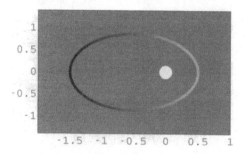

Recalling that red means fast and blue slow, we see that the planet hits maximum speed when it is closest to the sun, as it must do because of Kepler's law, which states that the planet sweeps out equal areas in equal time.

We can illustrate Kepler's law more convincingly by drawing lines from the sun at the origin to the planet at equal time intervals. To do this we store the solution in orbit (via the SolutionName option) and then do some graphics programming to get the lines we want. Since SaveSolution (which is automatically turned on by the use of SolutionName) causes only the plotted variables to be saved, substituting a Range for t in orbit will create a pair: some **X**-data and some **Y**-data. Transposing gives us the pairs we want, and we can Map the pure function that draws a line from the origin to a point onto this set of points.

```
e = 0.5;
orbitImage = SecondOrderPlot[
   {x"[t] == -x[t] / (x[t]^2 + y[t]^2)^(3/2),
    y"[t] == -y[t] / (x[t]^2 + y[t]^2)^(3/2)},
   {x[t], y[t]}, {t, 0, 2 Pi}, {x, -2, 1}, {y, -1.4, 1.4},
   InitialValues -> {1 - e, 0, 0, Sqrt[(1 + e)/(1 - e)]},
   PlotStyle -> AbsoluteThickness[2], SolutionName -> "orbit",
   DisplayFunction -> Identity,
   Epilog -> {AbsolutePointSize[6], Point[{0, 0}]}];

Show[orbitImage, Graphics[Map[Line[{{0,0}, #}] &, Transpose[
      orbit /. t -> Range[0, 2 Pi - Pi/15, Pi/15]]] ],
   DisplayFunction -> $DisplayFunction];
```

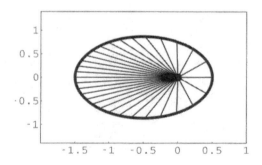

Another example of a second-order system arises from the flight of a thrown object, such as a ball. In the absence of air resistance, the model is quite simple. We assume that the mass equals 1; and in the following example the object is thrown form a height of 1.5 meters with initial velocity vector (20, 20); this is the same as saying the initial velocity is 28.28 meters per second, at an angle of 45°. The result is the familiar parabolic trajectory.

```
g = 9.8; m = 1;
SecondOrderPlot[{x"[t] == 0, y"[t] == -m g},
  {x[t], y[t]}, {t,0, 10}, {x, 0, 100}, {y, 0, 25},
  InitialValues -> {0, 1, 20, 20}];
```

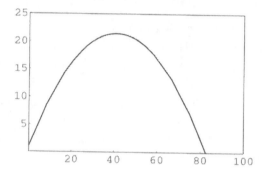

We can now add air resistance to the model. A typical assumption is that the resistance force varies with the square of the speed. Note that the angle of flight at time t is given by $\arctan(y'(t)/x'(t))$. The drag force is in the opposite direction (hence the minus signs) and splitting it into its x- and y-components yields the cosine and sine terms that arise. We use 0.002 as the resistance coefficient.

```
k = 0.002; g = 9.8;
SecondOrderPlot[{
x"[t] ==      - k (x'[t]^2 + y'[t]^2) Cos[ArcTan[y'[t]/x'[t]]],
y"[t] == -g - k (x'[t]^2 + y'[t]^2) Sin[ArcTan[y'[t]/x'[t]]]},
  {x[t], y[t]}, {t,0, 10}, {x, 0, 100}, {y, 0, 22},
  InitialValues -> {0, 1, 20, 20}];
```

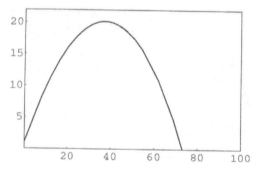

A much more complicated model of projectile motion is discussed in Chapter 13, where the flight of a discus is analyzed.

VisualDSolve and Differential Equations Modeling

Part 2

VisualDSolve and Differential Equations Modeling

Part 1 contains a complete description of the many functions and options of the `VisualDSolve` package. But, as with most software, a user requires detailed examples that show clearly how the package is used in specific, possibly quite complicated, situations. Part 2 contains a dozen such case studies. Some of these will be suitable for in-class presentations, while others, especially the later chapters, will be more appropriate for student projects.

Chapters 6 and 7 differ from the rest in that they are in a form suitable for a lab assignment in a course. These class-tested chapters include sequences of exercises, and the exposition is aimed at a student with only very little prior exposure to *Mathematica*. Chapter 6 is a notebook that serves as an introduction to *Mathematica*'s built-in functions for dealing with differential equations. Users should become familiar with the rudiments of the `DSolve` and `NDSolve` commands, and this notebook presents them in their most common forms. Chapter 7 is an introduction to the `VisualDSolve` command, centered around the problem of modeling a parachutist's downward flight.

Chapters 8, 9, and 10 present standard topics from the differential equations syllabus, using the `VisualDSolve` package to create appropriate and informative images and animations. The rest of Part 2 discusses diverse modeling problems, chosen to illustrate the many fascinating surprises that await the student of differential equations. Some of these models are straightforward (lead flow in the body, the flight of a discus), but others lead to interesting complications arising from issues of computation. For example, Chapters 11 and 15 address the important question of how one can attempt to determine if the computed solution is as accurate as it looks.

The final two chapters are somewhat more advanced, though still within the scope of a first course in differential equations. Chapter 16 illustrates an amazing theorem about the complexity inherent in a fairly simple model of a pendulum. It turns out that a complicated topological construction dating back to 1917—the Lakes of Wada—arises very naturally as basins of attraction for a forced, damped pendulum. Chapter 17 deals with more classical material: the behavior of a book tossed into the air. This chapter illustrates the predictive power that results when powerful software is combined with the classical modeling techniques of differential equations and physics. We discuss how to make a movie that correctly predicts what happens when a book is thrown into the air with an initial spin around its medium-length axis, and illustrate a classical theorem that relates the spinning book to a rolling egg.

Chapter 6

Differential Equations and Mathematica

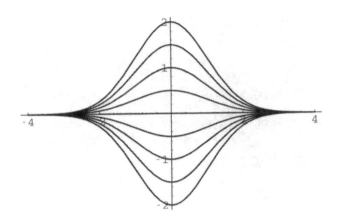

Overview

The `VisualDSolve` package tries to make it easy to generate images of solutions to differential equations. It is nevertheless important to have an understanding of *Mathematica*'s built-in functions for differential equations, and this chapter presents a sequence of commands and exercises to introduce those functions to students and instructors. The two main commands for solving a differential equation are `DSolve` and `NDSolve`, but the chapter also serves as a reminder about important *Mathematica* commands for integration, differentiation, and plotting.

6.1 ▪ Rules of Solving and DSolving Equations

Recall two important points about solving an equation: equality is denoted by ==; and the result is returned as a replacement rule, or rules.

```
quadraticSoln = Solve[x^2 - 3 x == 13, x]
```

$$\left\{\left\{x \to \frac{3 - \text{Sqrt}[61]}{2}\right\}, \left\{x \to \frac{3 + \text{Sqrt}[61]}{2}\right\}\right\}$$

The rules can be used to "plug in"; they cause x to be replaced by the right side of the rule(s). Simplify is used to simplify the result.

```
Simplify[x^2 - 3 x /. quadraticSoln]
{13, 13}
```

If you want to put the actual solutions in a home, do so as follows.

```
aHome = x /. quadraticSoln
```

$$\{\frac{3 - \text{Sqrt}[61]}{2}, \frac{3 + \text{Sqrt}[61]}{2}\}$$

Exercise 1: Try Solve on a variety of cubics, quartics, and quintics. Draw a conclusion about *Mathematica's* apparent abilities with regard to these types of equations.

Differential equations are attacked similarly by DSolve, though the syntax is a little more finicky. The unknown (a function!) is written as x[t] (or y[t] or foxes[t]), the unknown function gets listed after the equation, and the dependent variable is the third argument.

```
DSolve[x'[t] == t^2, x[t], t]
```

$$\{\{x[t] \rightarrow \frac{t^3}{3} + C[1]\}\}$$

Again, we get a rule, but it involves a constant, called C[1]. You could use C[1] -> C if you prefer plain old C. As with the quadratic equation example, using x[t] /. gets us the actual solution (% refers to the preceding output.)

```
DSolve[x'[t] == t^2, x[t], t] /. C[1] -> C
```

$$\{\{x[t] \rightarrow C + \frac{t^3}{3}\}\}$$

```
solution = x[t] /. %
```

$$\{C + \frac{t^3}{3}\}$$

Note the extra braces; they can be eliminated by using First[%] instead of just %. There are several important variations. Most often we will have an initial condition, such as $x(0) = 1$. That gets added to the equation like so:

```
DSolve[{x'[t] == t^2, x[0] == 1}, x[t], t]
```

$$\{\{x[t] \rightarrow 1 + \frac{t^3}{3}\}\}$$

```
solution = x[t] /. First[%]
```

$$1 + \frac{t^3}{3}$$

Let's differentiate as a check. You could just look at D[solution, t] and check that t^2 pops out. But let's be more general and plug the solution into the equation. Because solution is an expression in t (not a function, like Sin or Cos), we have to use the D operator to differentiate it (not just ').

```
(x'[t] == t^2)  /.  x'[t] -> D[solution, t]
True
```

Reminder about D[]

D[*expr*, v] takes the derivative of *expr* with respect to the variable v. This operation is very simple—the rules are well known to millions of teenagers the world over—so there is no need to check the results. Computers can handle even complicated expressions with no difficulty.

```
D[Tan[Sqrt[t] E ^ Sin[1/t]], t]
```

$$\left(\frac{E^{Sin[1/t]}}{2\ Sqrt[t]} - \frac{E^{Sin[1/t]}\ Cos[\frac{1}{t}]}{t^{3/2}}\right)\ Sec[E^{Sin[1/t]}\ Sqrt[t]]^2$$

If f is a function that you define as follows, then f'[x] and f'[3.2] work.

```
f[x_] := Sin[x^2]
{f'[r], f'[3.2]}
```

$$\{2\ r\ Cos[r^2],\ -4.38852\}$$

Of course, because x does not appear on the right side of $dx/dt = t^2$, this differential equation is nothing more than a simple integration problem, and could be attacked that way too (but then the necessary constants will not be shown).

```
Integrate[t^2, t]
```

$$\frac{t^3}{3}$$

Exercise 2: Fool around with `Integrate` to remind yourself of its power and limitations. Recall that most complicated functions do not have an elementary antiderivative. Sometimes, though, a new named function comes into play, analogous to the way logarithms show up when $1/t$ is integrated.

```
Integrate[Sin[t] * (1 + t^2), t]
```

```
Integrate[E^(-t^2), t]
```

```
Integrate[1/Log[t], t]
```

```
Integrate[Tan[t]/(1 + Sin[t]), t]
```

```
Integrate[Sin[t] (1 + Sqrt[t]), t]
```

```
Integrate[t^4 (1 - t)^4 / (1 + t^2), t]
```

The last example is somewhat amazing. To see why, evaluate the answer at 1 and 0 and subtract. Then plot the integrand (the function being integrated) of the previous example over the interval [0, 1]. What is unusual about the area defined by the plot? Use the `Plot` command. To remind yourself how `Plot` works, use `?Plot`.

Exercise 3: A request for `Integrate[Sqrt[Sin[t]], t]` yields no useful answer; this is because there is no elementary antiderivative. Suppose you really need to know the definite integral of this function between 0 and 2.5. How would you go about finding it?

6.2 ▇ Changing Politics: *x* Moves to the Right

Much more important is the case where x appears on the right side of a differential equation. We start with the simplest possible instance, which should be familiar to you.

```
DSolve[x'[t] == x[t], x[t], t]
```

$$\{\{x[t] \to E^t \; C[1]\}\}$$

We know this answer is correct. Here is a more complicated differential equation.

```
eqn = (x'[t] == x[t] - t^2);
DSolve[eqn, x[t], t]
```

$$\{\{x[t] \rightarrow 2 + 2\ t + t^2 + E^t\ C[1]\}\}$$

```
solution = x[t] /. First[%]
```

$$2 + 2\ t + t^2 + E^t\ C[1]$$

It is *always* wise to check; we can do so by comparing left and right sides.

```
{D[solution, t], solution - t^2}
```

$$\{2 + 2\ t + E^t\ C[1],\ 2 + 2\ t + E^t\ C[1]\}$$

Good. We can also be a little more abstract as follows; note that here we have to replace both `x[t]` and `x'[t]`.

```
eqn /. {x[t] -> solution, x'[t] -> D[solution, t]}
True
```

Exercise 4: Repeat the above sequence on the following differential equations, given in standard mathematical notation. Things will not always go as smoothly as in the preceding examples. Comment on the ways things can go wrong (there are several different things that can happen), and comment also on any patterns you notice. Do try variations on these equations.

In some of the cases, *Mathematica* will be unable to solve the problem. In at least one of these unsolvable cases, you can solve the differential equation by hand! Discover such a case and solve it.

(a) $x' = 3x + 15$

(b) $x' = x - t^2 - \sin t$

(c) $x' = x^2 - t$

(d) $x' = x - t^3$

(e) $x' = tx^2$

(f) $x' = |t|$ (absolute value is Abs [ ])

(g) $x' = |x|$

(h) $x' = \sin(x + t)$

Experiment with some equations of your own choosing, trying to find some patterns. Try to make some general statements about some classes of differential equations (with x appearing on the right side) that seem to be solvable symbolically. Check out some of the exercises in a text on solving differential equations and see if `DSolve` fails to handle any of them.

Our next goal is to produce a graphic that shows several solutions corresponding to different initial values. Here's a differential equation that `DSolve` can handle.

```
DSolve[y'[t] == -t y[t], y[t], t]
```

$$\{\{y[t] \rightarrow \frac{C[1]}{E^{t^2/2}}\}\}$$

Here is how one specifies initial conditions; basically all this does is find the right value of the constant `C[1]` for the given initial condition.

```
DSolve[{y'[t] == -t y[t], y[0] == 0.5}, y[t], t]
```

$$\{\{y[t] \rightarrow \frac{0.5}{E^{t^2/2}}\}\}$$

We may as well be more abstract and set the initial value to be `y0`. At the same time, we use `/.` to get the actual expression for the solution, and `First` to eliminate some braces.

```
y[t] /. First[
   DSolve[{y'[t] == - t y[t], y[0] == y0}, y[t], t]]
```

$$\frac{y0}{E^{t^2/2}}$$

```
soln[y0_] := y0 / Exp[t^2/2]
```

The `VisualDSolve` package makes this somewhat easier, but nevertheless, here is how one would generate several solution plots from scratch. The use of `Evaluate` is important since we want to turn the `Table` object into a list of functions before `Plot` sees it.

```
Plot[Evaluate[Table[soln[y0], {y0, -2, 2, 0.5}]], {t, -4, 4}];
```

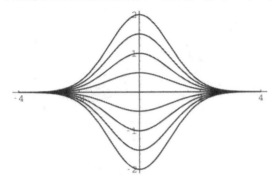

Exercise 5: Choose some of the differential equations in Exercise 4 (at least two) and follow the procedure just presented to get a nice picture of the family of solutions. You might have to experiment a bit with the choice of initial values. And notice that sometimes the solution comes out in more than one form (perhaps a square root with both positive and negative signs, or 0 treated as a special case).

By far the majority of interesting differential equations are *not* solvable symbolically. We will therefore have to understand how to attack those problems numerically, and to estimate the reliability of the results.

6.3   NDSolve

Many equations cannot be solved in terms of a nice formula ("closed form"), so it is important to have an idea of how *Mathematica* can be used to solve differential equations by numerical methods. Just as NIntegrate is the command for numerical integration, NDSolve is the numerical version of DSolve. Here is a quick introduction. NDSolve provides *numerical* solutions of differential equations.

```
s = NDSolve[{x'[t] == x[t]^2 - t, x[0] == 1}, x[t], {t, 0, 1}]
{{x[t] -> InterpolatingFunction[{0., 1.}, <>][t]}}
```

The result is a rule involving an interpolating function; such a function connects the dots, where the dots are points that are on the solution curve, or very close to it. An interpolating function can be plotted. First we give it a name. One way to do this is to go into the preceding output four levels deep and extract the actual function (recall that A[[i]] is the *i*th entry of a list A. A[[i, j]] is the *j*th entry of the *i*th entry. Moreover, things like rules have parts too: the second part of x[t] -> foobar is foobar. Short causes an abbreviated version of the output to be shown.

```
Short[solution1 = s[[1, 1, 2, 0]]]
InterpolatingFunction[{0., 1.}, <>]
```

This is an actual function, like sine or cosine, so we can plot it.

```
Plot[solution1[t], {t, 0, 1}];
```

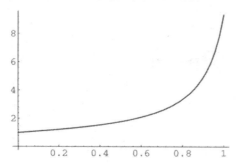

Alternatively, one can use the outputted rule to make a replacement; then we will get an expression in t.

```
solution2 = x[t] /. First[s]
InterpolatingFunction[{0., 1.}, <>][t]
```

With the solution in hand we can, for example, use FindRoot to find out when the solution takes on a particular value. We use the function form (solution1) and give the starting condition in the form of an interval.

```
FindRoot[solution1[t] == 4.55, {t, {0.6, 1}}]
{t -> 0.884111}
```

As usual, we check the answer.

```
solution1[0.884111]
4.55
```

```
solution2 /. t -> 0,884111
4.55
```

NOTE: If you use VisualDSolve and set the SaveSolution option to True then the solution is saved in a variable called Solution. However, it is saved as an expression (InterpolatingFunction[t]) rather than just a function. Thus Solution /. t -> 13 is the way to see specific values. Plot[Solution, ...] is how to plot it. And FindRoot[Solution, ...] is how to use FindRoot on it.

Exercise 6: Use NDSolve to obtain a solution to $dv/dt = -9.8 - 0.2\sqrt{1 + v^2}$, with initial condition $v(0) = 0$. Plot the solution. Determine the value of t so that $v(t) = -200$.

Chapter 7

Some Parachute Experiments

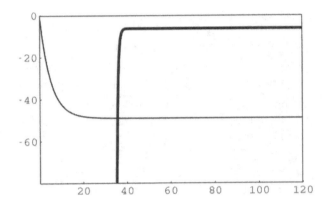

Overview

This chapter presents a lab for students that makes use of VisualDSolve to model a real-world problem. This lab serves both as a general introduction to the VisualDSolve package, as well as a specific introduction to some subtle aspects of a realistic problem about a parachutist. The main points of this example were presented at a differential equations workshop by Dieter Armbruster and Eric Kostelich of Arizona State University, authors of [AK].

7.1 ▨ VisualDSolve

The VisualDSolve package allows one to generate solutions to differential equations in a convenient and flexible way. To load the package, evaluate the following cell.

```
Needs["VisualDSolve`"]
```

The package contains many, many functions. The most basic function is called VisualDSolve.

```
?VisualDSolve
```

VisualDSolve[equation, {t, tmin, tmax}, {x, xmin, xmax}, (opts)]
    creates an image that shows the solutions to a first-order ODE
    superimposed on a field of slope lines. There are LOTS of
    options! And all of the options for Graphics, Plot, ContourPlot,
    and NDSolve may be used as well. The equation must be input in
    the standard form such as x'[t] + x[t] == t + x[t]^2. Note also
    that the first iterator defines the independent variable and the
    other iterator defines the dependent variable.

VisualDSolve takes three arguments: an equation in terms of functions of the
independent variable (we will assume it is t, but any name can be used; likewise we
call the dependent variable x), a *t*-iterator, and an *x*-iterator. The two iterators define
the viewing window. In order to see solutions, one must specify some initial values via
the InitialValues option. This can be a specific *t-x* pair, a set of such pairs, or a
setting such as Grid[3]. The ShowInitialValues option causes the initial values
to be shown as dots.

```
VisualDSolve[x'[t] == x[t], {t, 0, 3}, {x, 0, 3},
    InitialValues -> {0, 1}];
```

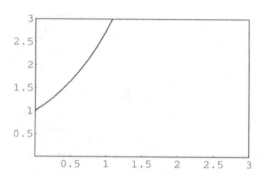

```
VisualDSolve[x'[t] == x[t], {t, 0, 3}, {x, 0, 3},
    InitialValues -> Grid[3], ShowInitialValues -> True];
```

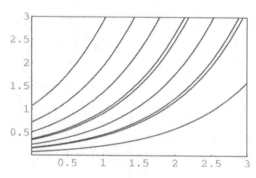

If we wish to see many solutions, then the `Table` command is an efficient way to generate the initial values.

```
Table[{0, i}, {i, -3, 3, 0.5}]

{{0, -3}, {0, -2.5}, {0, -2.}, {0, -1.5}, {0, -1.},
  {0, -0.5}, {0, 0.}, {0, 0.5}, {0, 1.}, {0, 1.5},
  {0, 2.}, {0, 2.5}, {0, 3.}}


VisualDSolve[x'[t] == x[t], {t, -0.1, 3}, {x, -6, 6},
  InitialValues -> Table[{0, i}, {i, -3, 3, 0.5}],
  ShowInitialValues -> True];
```

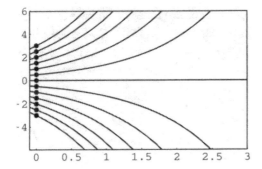

We can also ask that a symbolic approach be tried. In this case it works and the general solution is displayed.

```
VisualDSolve[x'[t] == x[t], {t, 0, 3}, {x, -3, 3},
  ShowInitialValues -> True,
  SymbolicSolution -> True,
  InitialValues -> Table[{1, x}, {x, -2, 2, 0.5}]]
```

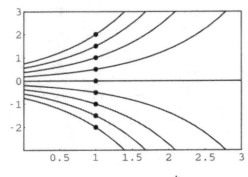

```
{-Graphics-,- {{x[t] -> C E^t}}}
```

There are a gazillion options (`Options[VisualDSolve]` will show them). Chapter 1 gives the details.

Exercise 1: Read the first few sections of Chapter 1 and try out various things. In particular, take some of the examples from a differential equation textbook and use VisualDSolve to create instructive images. Explain anything you see in your graphs that looks interesting or intriguing or surprising.

7.2 ▮ Modeling a Parachutist

An object falling under the influence of gravity feels an acceleration of −9.8 meters/second/second (negative because we take "up" as being the positive direction). This simplification ignores two very important things: (1) air resistance, and (2) ground resistance. (We can ignore the slight variation in g; 9.8 is a reasonable value near the earth's surface.) Standard notation here is s for distance (height above the ground; we assume that our parachutist starts 2000 meters up: $s(0) = 2000$), v for velocity (or speed, but be very careful of the sign), and a for acceleration. The fact that acceleration is constant (we take g to be +9.8, so $a(t) = -g$) leads to a simple differential equation for v: $v'(t) = -g$, with $v(0) = 0$.

```
g = 9.8;
VisualDSolve[v'[t] == -g, {t, 0, 100}, {v, -1000, 0},
    InitialValues -> {0, 0},
    AxesLabel -> {"t (sec)", "v (meters per second)"}];
```

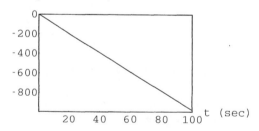

Note that $v' = -g$ can be viewed as a second-order equation, $s''(t) = -g$; but for now let's see what we can learn about speed by using first-order equations.

Of course, we did not need a computer to solve this simple differential equation. Look at the output and imagine what it means for a chuteless parachutist (useful facts: 0.44 meters per second = 1 mph; 10 meters per second = 22.3 mph).

Of course, the two factors we ignored are actually pretty important. The ground would put a quick stop to a parachutist's acceleration. More subtle is the effect of air resistance. In fact, people have survived a chute's failure to open; this would be impossible without air resistance. A reasonable model is that the retarding force of the air goes up with the speed of the falling object. For a jumper of average weight, 0.2 is a reasonable coefficient, so we have the differential equation: $v'(t) = -g - 0.2v(t)$. Note that air resistance is a force, and so affects acceleration.

```
VisualDSolve[v'[t] == -g - 0.2 v[t],
   {t, 0, 50}, {v, -80, 0},
   InitialValues -> {0, 0},
   AxesLabel -> {"t (sec)", "v (mps)"}];
```

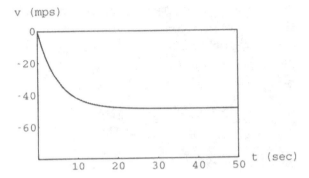

Exercise 2:

(a) Solve $v'(t) = -9.8 - 0.2v(t)$ symbolically. Do it by hand (standard calculus techniques will work: separation of variables). Then try *Mathematica* using either DSolve (as explained in Chapter 6) or VisualDSolve with the SymbolicSolution option set to True.

(b) Check the symbolic solution by differentiating, using D[..., t].

(c) Explain why the exact solution proves that the apparent asymptote really is an asymptote. What is the exact limiting value?

(d) Give a physical interpretation of the existence of the limiting value. Hint: Drop a small piece of paper from a height of 6 feet and ponder its speed.

Exercise 3:

Suppose that, at the instant the parachutist jumps, the plane was diving at a good rate of speed (say, 200 mps in the vertical direction). Repeat the preceding computations so see the effect on v. Suppose the plane was climbing? Use VisualDSolve to create a single image that shows several of these situations.

Exercise 4:

(a) What is t_1, the time it takes the parachutist to hit the ground? Note that $s(t_1)$ is 2000 less the distance fallen to time t_1, and the latter—the distance fallen—is the integral of $v(t)$ from 0 to t_1 (actually the negative of this, because the integral will be negative since v is negative). You have $v(t)$ in closed form from Exercise 2; integrate it and try, by hook or by crook, to find a value of t_1 for which this integral is −2000, which will yield $s = 0$.

(b) When will the parachutist be 500 meters above the ground? This value will be needed in the next section when we try to understand what happens when the parachute opens at the 500-meter level.

7.3   ■   Parachute to the Rescue

We now wish to understand what happens to the velocity when the chute opens as it should. Assume that the chute is opened 500 meters from the ground, which, you should have found from the last exercise, occurs after 35.6 seconds of free-fall. We take the coefficient of drag for a person-and-chute to be 1.56. This means that the differential equation involves two cases:

$$v'(t) \;=\; -g - 0.20v(t) \qquad \text{if } t < 35.6$$
$$v'(t) \;=\; -g - 1.56v(t) \qquad \text{if } t \geq 35.6$$

This can be solved in closed form (think about that), but instead of puzzling that out it is simpler to use numerical methods. Since we can feed an If-statement to VisualDSolve, it is easy to see the solution.

```
VisualDSolve[v'[t] == -g - If[t < 35.6, 0.2, 1.56] * v[t],
    {t, 0, 120}, {v, -80, 0},
    InitialValues -> {0, 0},
    AxesLabel -> {"t (sec)", "v (mps)"}];
```

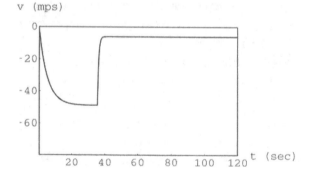

Exercise 5: Interpret the preceding graph. Does it look like an accurate portrayal of what might happen to a parachutist? Is the ground-strike-speed reasonable?

In fact, the preceding solution is a composite of two solutions. The discussion that follows generates the two solutions to different equations that combine to yield the parachutist's speed profile shown above.

```
VisualDSolve[v'[t] == -g - 0.2 * v[t],
   {t, 0, 120}, {v, -80, 0},
   InitialValues -> {0, 0}, SymbolicSolution -> True,
   AxesLabel -> {"t (sec)", "v (mps)"}]
```

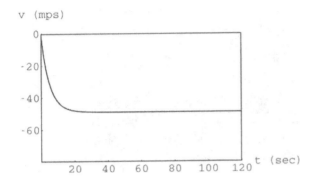

```
                                          C
{-Graphics-, {{v[t] -> -49. + ---------}}}
                                      0.2 t
                                    E
```

Because $v(0) = 0$, the constant in the symbolic solution is 49. We use this to find the speed when the chute opens.

```
v[t] /. %[[2]] /. {t -> 35.6, C -> 49}
```

```
{-48.9604}
```

We now generate a composite image showing the two paths: one for a human in free-fall, the other for a parachutist slowing down.

```
Show[
   VisualDSolve[v'[t] == -g - 1.56 * v[t],
      {t, 0, 120}, {v, -80, 0},
      InitialValues -> {35.6, -48.9604},
      PlotStyle -> AbsoluteThickness[2],
      DisplayFunction -> Identity],
   VisualDSolve[v'[t] == -g - 0.2 * v[t],
      {t, 0, 120}, {v, -80, 0},
      InitialValues -> {0, 0}, DisplayFunction -> Identity],
DisplayFunction -> $DisplayFunction];
```

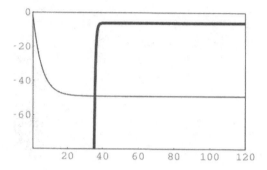

Exercise 6: For all the parameters as given above, compute the total time it will take before the parachutist hits the ground. This is quite similar to Exercise 4, and we can again make use of the fact that we have a symbolic expression for $v(t)$.

7.4 ■ Infinite Jerk Strikes Again: Dead Body Hits Ground!

The behavior predicted by our model is only partly reasonable. The speed at which the parachutist hits the ground seems quite appropriate. But, unfortunately, under our assumptions he or she would be dead before that happened! This is because our model assumed that the parachute went from fully closed to fully open in an instant. This would cause a tremendous strain on the body, as the velocity is changed from over 100 miles per hour to about 13 miles per hour in a very short time interval! In fact, we can see from our model that the deceleration at the instant the parachute opens is 66 meters per second per second, or almost 7 *gs*. We get this by simply letting v be -49 into the defining expression for dv/dt.

```
-g - 1.56 * (-49)

66.64


% / g

6.8
```

This number by itself does not mean instant death, but the fact that the acceleration changes in an instant means that the jerk felt by the falling body is infinite. This can be seen in the sudden change in velocity that occurs at the sharp bend in the figures of section 7.3. Incidentally, "jerk" is, in fact, a technical term referring to the derivative of acceleration. For an analogy think of a human who jumps off a cliff with a rope tied to a harness (like bungee jumping). If the rope had no stretch in it, the human's internal organs would not survive the jerk.

An obvious way to improve the model is to allow some time, say 3 seconds, for the parachute to open. The simplest way to do this is to incorporate a linear

Color Plates

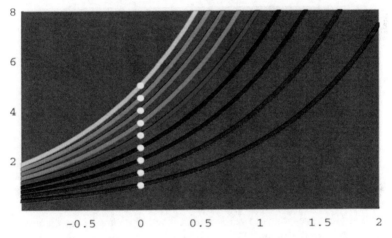

Color Plate 1. The **Rainbow** option causes the solutions corresponding to different initial values to be shown in different colors.

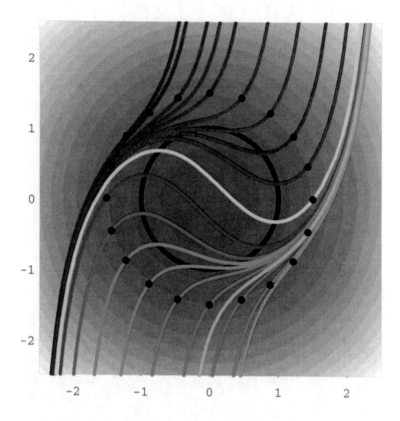

Color Plate 2. Combining colored solution curves with a background that shades regions according to the slope steepness gives a striking image. In this example, which shows solutions to $x' = x^2 + t^2 - 1$, the black circle corresponds to zero slope: those points are where the solutions have zero derivative. The black points are the initial values.

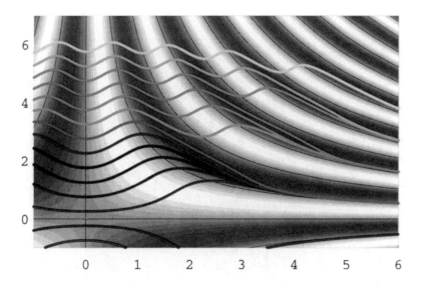

Color Plate 3. The colored curves are solutions to $x'(t) = \sin(t\,x)$. The gray background measures the steepness and the thin black curves are nullclines: they show the regions where the slope is zero.

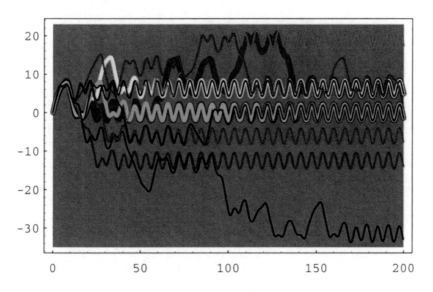

Color Plate 4. Eleven solutions (graphs of x vs. t) corresponding to different initial velocities ($x'(0)$) for a driven pendulum. It is very hard to predict which of the many periodic attractors a given solution will end up in. See page 56.

Concentrations

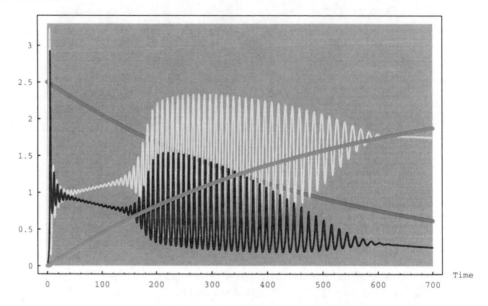

Color Plate 5. Concentrations of four chemicals in an autocatalator chemical reaction. This image was generated by **SystemSolutionPlot**. See page 59.

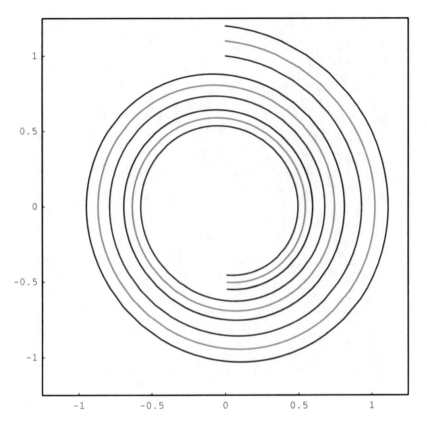

Color Plate 6. Three orbits for the system $x' = y$, $y' = -x - y/10$. See page 66.

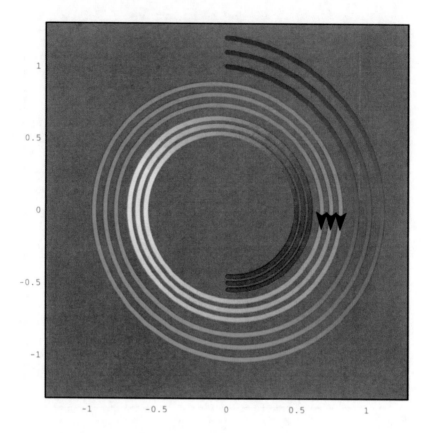

Color Plate 7. Three orbits graphed using the `ColorParametricPlot` setting for the `ParametricPlotFunction` option to `PhasePlot`. This colors each orbit in a hue that changes (from blue to red) as time increases. See page 67.

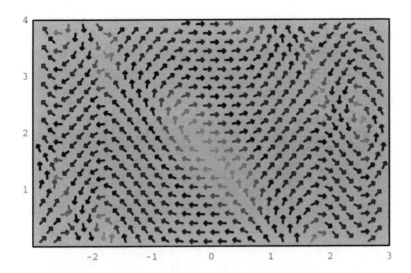

Color Plate 8. `PhasePlot` can draw vector fields so that each arrow is colored according to its true length, blue representing short vectors and red, long ones. In this case, however, the actual lengths displayed have been equalized. See page 72.

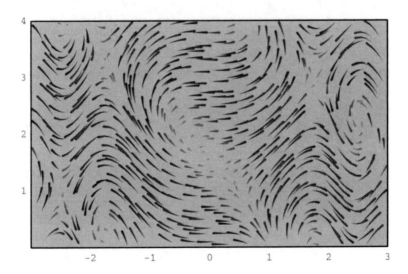

Color Plate 9. `PhasePlot` can draw flow fields using curvy fish shapes to represent the flow. In this example, the colors correspond to lengths, but the fish lengths have been slightly modified using the **FieldLogScale** option so that they present an informative image. If actual lengths were shown, the small ones would be too small for their direction to be evident. See page 72.

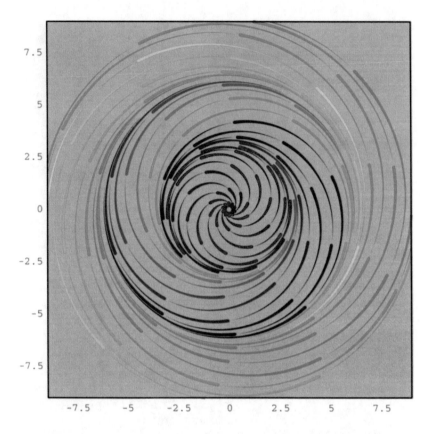

Color Plate 10. When **FlowParametricPlot** is used to draw orbits they are drawn as a sequence of linked fish. In this example **Rainbow** is turned on so the 30 different initial conditions yield differently colored trajectories. This example is derived from a differential equation described using polar cordinates by $r' = \sin r$, $q' = 1$. See page 74.

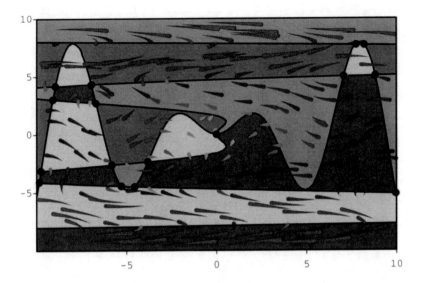

Color Plate 11. The shaded nullcline regions for a two-dimensional autonomous system show where the underlying vector field is pointing northeast, northwest, southeast, or southwest. In this example the flow field is superimposed, colored according to the strength of the flow. The black curves are the nullcline curves; thus the flow at points on those curves is either vertical ($x' = 0$) or horizontal ($y' = 0$). See page 84.

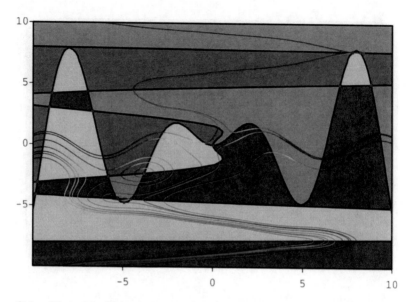

Color Plate 12. This shows some orbits for the system in plate 11. The orbits are colored according to increasing time. Again, the shaded null-cline regions add a lot of structure to the phase plane view. The intersections of the nullclines are the equilibrium points (shown in red), which **PhasePlot** can find automatically. See page 84.

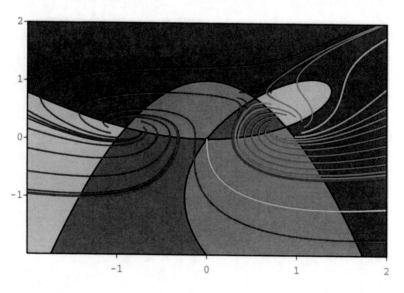

Color Plate 13. The system $x' = x^3 - 4xy + 2y^2 + y^3$, $y' = -1 + x^2 + y$ has the curious feature that the x-nullcline (the curve corresponding to $x' = 0$) crosses itself. This is not an equilibrium point. There are three equilibirum points, shown in white. See page 86.

Color Plate 14. The use of shaded nullcline regions and flow fields can often provide enough information that the actual orbits are almost redundant. In this example one can clearly imagine what all the orbits will look like. See page 86.

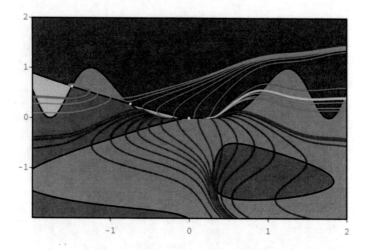

Color Plate 15. The nullclines can sometimes yield closed loops, as in this example. Of course, the flow within such a loop will be in a single general direction, northeast in this case. See page 87.

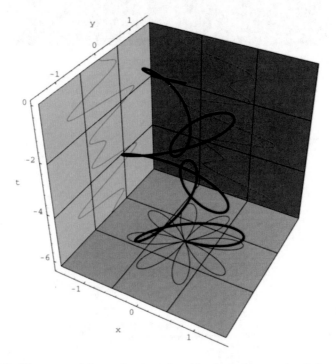

Color Plate 16. The parametric curve given in polar form by $r = \sin 4\theta$ is an 8-leafed rose. When **ProjectionPlot3D** is used to plot such a curve, the result is the actual graph in 3-space. The usual projection is shown on the horizontal plane, while the other two planes show the x and y cordinates as individual functions. See page 91.

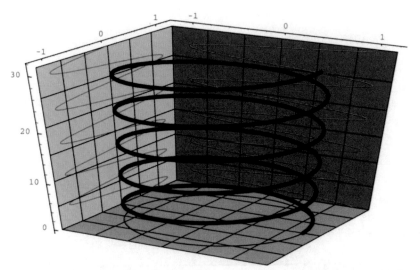

Color Plate 17. When **PhasePlot** is called with three viewing iterators, **Projection-Plot3D** is used to plot the three-dimensional space curve. The image here shows the result for the simple system: $x' = -y$, $y' = x$. The orbit in the plane is a circle, but the full graph is a helix. See page 91.

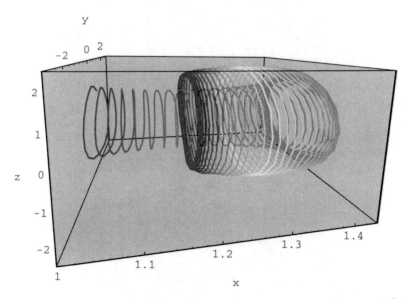

Color Plate 18. **PhasePlot** can generate the space curve corresponding to a trajectory of a three-dimensional system. The example shown here, which comes from a system that describes a type of electrical circuit, has been colored so that the changing colors corresponds to increasing time. The mushroom shape of the orbit is clearly visible. See page 95.

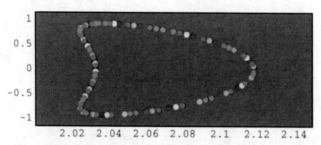

Color Plate 19. A Poincaré section shows points on an orbit at regular intervals. This can often provide easier-to-read images than a view of the entire orbit. In this example, the points are colored so that the hue changes as time increases. See page 96.

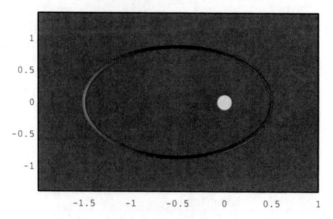

Color Plate 20. A planet orbiting a sun. The coloring is according to the speed of the planet (red is fast, blue is slow), and so illustrates Kepler's law that the radial vector from the sun sweeps out equal areas in equal times. See page 108.

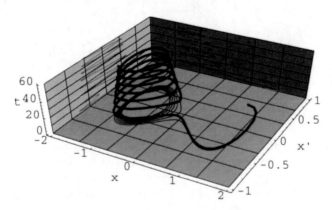

Color Plate 21. A view of the true graph of a stable orbit of the Duffing equation, with the three projections shown on the walls. See page 204..

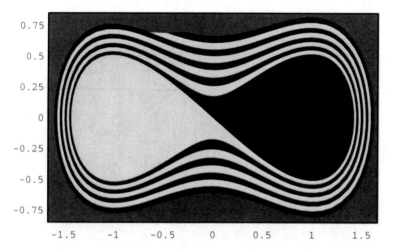

Color Plate 22. The two basins of attraction for a Duffing equation with small damping constant and no forcing. Initial conditions in the blue region converge to the rightmost equilibrium; yellow starting points go to $(-1, 0)$. There is also an equilibrium point at $(0, 0)$. This image was generated as two **Polygon** objects, using the data inherent in two special solutions to the equation. See page 207.

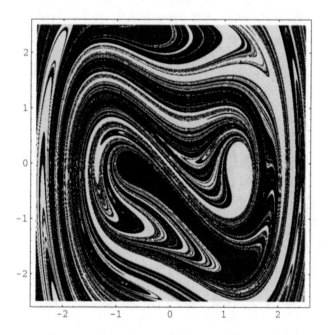

Color Plate 23. The basins of attraction for a Duffing equation with a modest amount of forcing. There are no equilibrium points, but trajectories tend to settle in one of two areas. The color coding indicates which initial conditions lead to which areas. This image is based on a 1000×1000 grid, and time was run out to 16π seconds. See page 217.

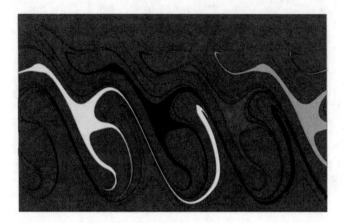

Color Plate 24. Lakes of Wada. The trajectories of the differential equation governing the motion of a damped pendulum subject to a periodic external force eventually settle into oscillatory behavior. In this image, each colored region represents initial points that settle into the same region under this oscillatory behavior. See page 235.

Color Plate 25. In physical systems solution curves often arise as the intersection of two surfaces. The example shown here presents the curve describing the rotation axis of a thrown book as the intersection of two ellipses, one arising from conservation of energy and the other from conservation of angular momentum. See page 248.

function into the differential equation. The code that follows does this, using a Which construction to make the drag coefficient change uniformly form 0.2 to 1.56 over a three-second period.

```
dt = 3;
VisualDSolve[v'[t] == -g - v[t] * Which[
   t < 35.6,        0.2,
   t < 35.6 + dt,   0.2 + 1.36 (t - 35.6)/dt,
   t >= 35.6 + dt, 1.56],
   {t, 0, 120}, {v, -60, 0},
   InitialValues -> {0, 0}, SolutionName -> "velocity",
   AxesLabel -> {"t (sec)", "v (mps)"}];
```

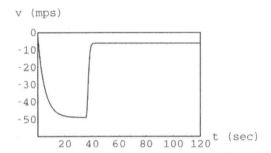

We can now figure out the g-forces by substituting the velocity function (an InterpolatingFunction object that represents the solution and was saved because the SolutionName option was turned on) that we have stored in velocity into the expression for dv/dt in the defining differential equation. That will give dv/dt, or acceleration.

```
acceleration = -g - velocity * Which[
   t < 35.6, 0.2, t < 35.6 + dt,
   0.2 + 1.36 (t - 35.6)/dt, t >= 35.6 + dt, 1.56];

Plot[acceleration, {t, 34, 42}, PlotRange -> {-1, 25},
   AxesOrigin -> {34, -1}, AxesLabel -> {"time", "acceleration"}];

Plot[acceleration, {t, 35, 36}, PlotRange -> {-1, 15},
   AxesOrigin -> {35, -1}, AxesLabel -> {"time", "acceleration"}];
```

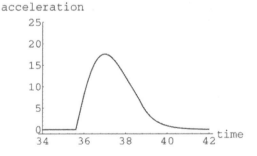

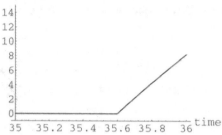

This shows quite modest acceleration (about 2 *gs*). The close-up view shows that that the jerk is now under control, since the slope of the acceleration curve is modest. We can see the infinite jerk by redoing the computation with the time interval (dt) set to 0, which is exactly the situation of our first model.

```
dt = 0;
VisualDSolve[v'[t] == -g - v[t] * Which[
   t < 35.6,        0.2,
   t < 35.6 + dt,   0.2 + 1.36 (t - 35.6)/dt,
   t >= 35.6 + dt, 1.56], {t, 0, 120}, {v, -60, 0},
   InitialValues -> {0, 0}, SolutionName -> "velocity",
   DisplayFunction -> Identity];

acceleration = -g - velocity * Which[
   t < 35.6, 0.2, t < 35.6 + dt,
   0.2 + 1.36 (t - 35.6)/dt, t >= 35.6 + dt, 1.56];

Plot[acceleration, {t, 34, 42}, PlotRange -> {-5, 70},
   AxesOrigin -> {34, -5},
   AxesLabel -> {"time", "acceleration"}];
```

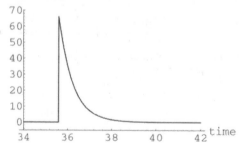

Chapter 8

Linear Systems

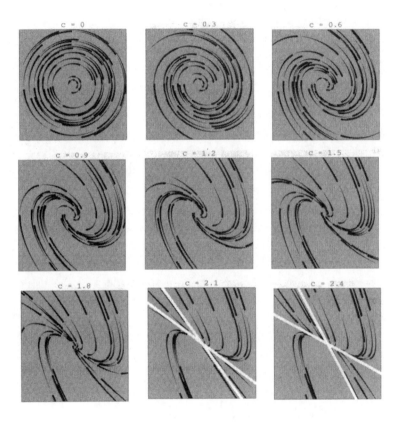

Overview

Autonomous linear systems are essentially the only large class of differential equations for which a fairly complete theory is available to solve the systems in general. These systems can be written compactly in the form $\mathbf{X}'(t) = \mathbf{A}\mathbf{X}(t)$, where $\mathbf{X}(t) = (x_1(t), x_2(t), \ldots, x_n(t))$. They are important not only because they can be solved, but also because many applications can be closely modeled with linear systems. Further, an important technique in the study of nonlinear systems involves the study of equilibrium points and they are classified by linearizing the system at the equilibrium point.

8.1 ▆▆ A Comprehensive View of the Two-Dimensional Case

The theory of autonomous linear differential systems is essentially a direct application of linear algebra. Solutions are found by analyzing eigenvalues and eigenspaces of the matrix defining the system. If $\mathbf{v}$ is an eigenvector for $\mathbf{A}$ with eigenvalue λ, it is easy to verify that $e^{\lambda t}\mathbf{v}$ is a solution to the system $\mathbf{X}' = \mathbf{AX}$. For an n-dimensional system, if we are lucky enough to get n independent solutions in this manner, then we will be able to construct the general solution of the system. In the other cases it is still possible to construct the general solution using more sophisticated linear algebra techniques.

In the two-dimensional case, there are 13 distinct types of behavior. Twelve of the cases are shown on the next two page, where the white lines are the solutions to the system gotten from the cases where real eigenvalues were found with one-dimensional eigenspaces.

Six of the cases correspond to the existence of two distinct eigenvalues for the matrix; both real and positive, both real and negative, one positive and one negative, imaginary with positive real part, imaginary with negative real part, and imaginary with zero real part. These are in the last two rows of the array. The three possible cases with a 0 eigenvalue (the first row)—both 0, one positive, or one negative—are special in that the origin is not an isolated equilibrium point and thus cannot be described as stable or unstable. The second row shows three of the four possibilities where the matrix has only one eigenvalue. That eigenvalue is either positive or negative and has either a one- or two-dimensional eigenspace. The 13th case is the one-eigenvalue situation (example: $x' = -x$, $y' = -y$) with a single negative eigenvalue having a two-dimensional eigenspace; this results in a star node with the motion toward the center.

The code that follows produces the large graphics array showing the 12 cases.

```
Needs["VisualDSolve'"];

Show[GraphicsArray[Partition[PhasePlot[
    Thread[{x'[t], y'[t]} == (#[[1]] . {x[t], y[t]})],
    {x[t], y[t]}, {x, -3, 3}, {y, -3, 3}, FlowField -> True,
    FieldColor -> RandomGrayLevel,
    NullclineShading -> True, NullclinePlotPoints -> 10,
    AspectRatio -> Automatic, Ticks -> None,
    DisplayFunction -> Identity,
    Epilog -> {If[FreeQ[#[[1]]], 0],
        {Thickness[0.02], White,
        Line[{3 #[[2, 1]], - 3 #[[2, 1]]}],
        Line[{3 #[[2, 2]], - 3 #[[2, 2]]}]}], {}],
        AbsolutePointSize[5], Point[{0, 0}]}] & /@
    Map[({#, Eigenvectors[#]}) &, {
        {{0,.0}, {0, 0}}, {{1, 1}, {1, 1}}, {{-1, -1}, {-1, -1}},
        {{1, 0}, {0, 1}}, {{1, -4}, {1, 5}}, {{-1, -4}, {1, -5}},
        {{-1, -3}, {1, -5}}, {{1, -2}, {1, 4}}, {{1, 3}, {1, -1}},
        {{0, 1}, {-1, -1}}, {{0, 1}, {-1, 1}}, {{0, 1}, {-1, 0}}}],
    3]], DisplayFunction -> $DisplayFunction];
```

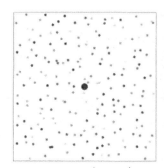

The zero system: $x' = 0$, $y' = 0$

Eigenvalues: 0 with 2-dim
   eigenspace

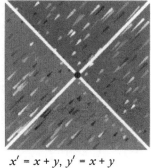

$x' = x + y$, $y' = x + y$

Eigenvalues: 0, 2

Motion away from 0-eigenspace

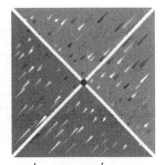

$x' = -x - y$, $y' = -x - y$

Eigenvalues: 0, -2

Motion toward 0-eigenspace

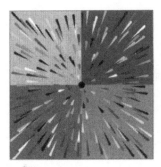

The zero system: $x' = x$, $y' = y$

Eigenvalues: $+1$ with 2-dim
   eigenspace

Unstable equilibrium

Proper node: $x' = x - 4y$,
   $y' = x + 5y$

Eigenvalues: $+3$ with 1-dim
   eigenspace

Unstable equilibrium

Proper node: $x' = -x - 4y$,
   $y' = -x - 5y$

Eigenvalues: -3 with 1-dim
   eigenspace

Asymptotically stable equilibrium

Improper node: $x' = -x - 3y$,
   $y' = x - 5y$

Eigenvalues: -4, -2

Asymptotically stable equilibrium

Improper node: $x' = x - 2y$,
   $y' = x + 4y$

Eigenvalues: 2, 3

Unstable equilibrium

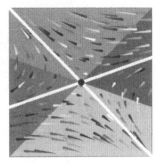

Saddle: $x' = x + 3y$, $y' = -x - y$

Eigenvalues: -2, $+2$

Unstable equilibrium

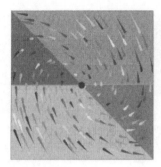 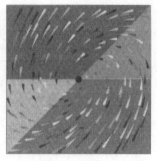 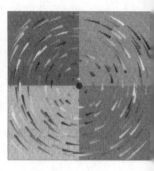

Improper node: $x' = y$,   Focus: $x' = y$, $y' = -x + y$ Center: $x' = y$, $y' = -x$

$y' = -x - y$ Eigenvalues: $0.5 \pm 0.87i$ Eigenvalues: $i, -1$

Eigenvalues: $-0.5 \pm 0.87i$ Unstable equilibrium   Stable equilibrium

Asymptotically stable equilibrium

8.2 ▮ Physical Application: Springs

In this section we consider a linear system that arises from Newton's law of motion and Hooke's law for the restoring force of a spring. Consider a mass suspended from a spring as illustrated.

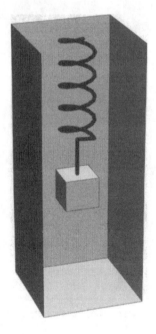

Newton's law of motion says that the rate of change of momentum of an object is equal to the sum of the forces acting on the object. Momentum is defined as mass times velocity. Newton's law, in mathematical terms is : $(d/dt)(mv) = F$, where m is the mass, v is the velocity, and F is the total force acting on the object. The most common case is to have a constant mass which reduces to the more familiar equation $F = ma$.

The force of gravity, mg, pulls down on the object where, g is the gravitational constant. There is also the spring acting on the object. We use Hooke's law which says that the force of the spring is proportional to how far the spring is stretched. This is approximately true for short displacements of a good spring. Ignoring other forces (such as air resistance) we then have:

$$m\frac{dv}{dt} = mg - kx$$

where x is the displacement of the spring from its rest position and k is Hooke's constant for the spring. Note that we are assuming that down is the positive direction.

Adding $dx/dt = v$ to the above equation yields a first-order system that will have a unique solution for given initial values of x and v. Note that $x = mg/k$, $v = 0$ is a constant solution to the system, known as an equilibrium solution. In fact, one could calculate k by locating this equilibrium solution for the apparatus and measuring the displacement x.

It is convenient to substitute $y = x - mg/k$ into the system; $y' = x' = v$ and so the system becomes

$$y' = v$$
$$v' = -\frac{k}{m}y$$

This gives a two-dimensional linear autonomous system. In *Mathematica* notation we write the system as follows.

```
springSys = {y'[t] == v[t], v'[t] == -(k/m) y[t]};
```

Mathematica will easily solve this system for any initial values. For example:

```
springSoln = DSolve[Join[springSys /.
   {k -> 1/3, m -> 1/3}, {y[0] == y0, v[0] == v0}], {y[t], v[t]}, t]
```

$$\left\{\left\{y[t] \to \left(\frac{I}{2}E^{-It} - \frac{I}{2}E^{It}\right)v0 + \left(\frac{E^{-It}}{2} + \frac{E^{It}}{2}\right)y0,\right.\right.$$

$$\left.\left.v[t] \to \left(\frac{E^{-It}}{2} + \frac{E^{It}}{2}\right)v0 + \left(\frac{-I}{2}E^{-It} + \frac{I}{2}E^{It}\right)y0\right\}\right\}$$

In general, DSolve will return complex-valued solutions. The usual way of getting the real parts of the solutions is to use ComplexExpand.

```
springSoln = ComplexExpand[{y[t], v[t]} /. First[springSoln]]
```

```
{y0 Cos[t] + v0 Sin[t], v0 Cos[t] - y0 Sin[t]}
```

SystemSolutionPlot provides a convenient way of looking at plots of the solutions for given initial values.

```
SystemSolutionPlot[springSys /. {k -> 1/3, m -> 1/3},
    {y[t], v[t]}, {t, 0, 10}, InitialValues -> {0, 0, 3}];
```

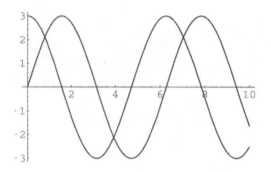

We can generate a movie for the motion of the spring for the above initial conditions. Only the first image from the animation is displayed here.

```
l = 6; b = 15;

Do[pos = springSoln[[1]] /. {y0 -> 0, v0 -> 3 , t -> s};
    springPts = Join[
        Table[{Sin[5 t], Cos[5 t], -(t (1 + pos)/l)}, {t, 0, 1, 0.1}],
            {{0, 0, -(1 + pos)},{0, 0, -(1 + pos + 2)}}]];
    Show[Graphics3D[{
            {Thickness[0.02], Line[springPts]},
            {GrayLevel[0.5], Cuboid[{1, 1, -(1 + pos + 2)},
                    {-1, -1, -(1 + pos + b/4)}]},
            {GrayLevel[0.25], Cuboid[{3, 3, 0}, {-3, -3, -b}]}}],
        Boxed -> False, ColorOutput -> GrayLevel,
        PlotRange -> {{-4, 4}, {-4, 2.9}, {0, -b - 1}},
        ViewPoint -> {1.32, 4.02, 1.71}],
    {s, 0, 23 Pi/12, Pi/12}]
```

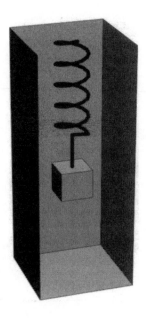

Since these systems are autonomous we can also view the orbits of the solutions in the phase space of the system. We now add a third force, damping, to our spring mass system. The simplest model for damping is to add a force proportional to the velocity of the system.

```
dampedSpringSystem = {y'[t] == v[t], v'[t] == -(k/m) y[t] - c v[t]};
```

To see the effect of damping, we can look at the phase portrait for increasing values of c.

```
inits = Table[
   {0, Random[Real, {-5, 5}], Random[Real, {-5, 5}]}, {20}];

Do[PhasePlot[dampedSpringSystem /. {k -> 1, m -> 1},
     {y[t], v[t]}, {t, -2, 2}, {y, -6, 6}, {v, -6, 6},
   InitialValues -> inits, StayInWindow -> True,
   ParametricPlotFunction -> FlowParametricPlot,
   NumberFish -> 4, Segments -> 20, AspectRatio -> Automatic,
   Ticks -> None, WindowShade -> Gray,
   PlotLabel -> StringForm["c = ``", c],
   Epilog -> If[c < 2, {},
        {AbsoluteThickness[2], White,
           Line[{{-6, -6 (-c + Sqrt[c^2 - 4])/2},
                 { 6,  6 (-c + Sqrt[c^2 - 4])/2}}],
          Line[{{-6, -6 (-c - Sqrt[c^2 - 4])/2},
                 { 6,  6 (-c - Sqrt[c^2 - 4])/2}}]}]],
     {c, 0, 2.7, 0.3}];
```

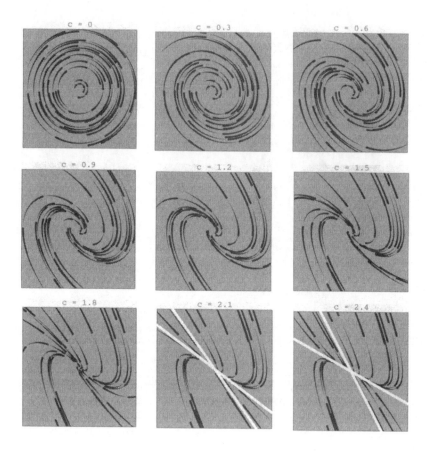

For large enough values of c, the eigenvalues are both real and negative. The two white lines in the last two images are the eigenvectors. In these cases the solutions tend toward zero without any oscillatory behavior, and the system is referred to as *overdamped.*

8.3 ■ A Four-Dimensional Example

We next consider a four-dimensional example arising from two masses interconnected with three springs.

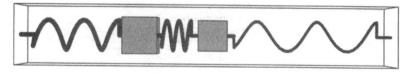

The equations are derived by considering the position of each mass separately and using Newton's and Hooke's laws as before. We assume masses of m_1 and m_2 and spring constants of k_1, k_2, and k_3; x_1 and x_2 are the positions of the masses measured from their equilibrium positions and v_1 and v_2 are their velocities. The system turns out to be:

```
X = {x1[t], v1[t], x2[t], v2[t]};
A = {{0, 1, 0, 0}, {-(k1 + k2)/m1, 0, k2/m1, 0},
     {0, 0, 0, 1}, {k2/m2, 0, -(k2 + k3)/m2, 0}};

twoMassSpring = Thread[D[X, t] == A . X]
```

$$\{x1'[t] == v1[t], \quad v1'[t] == -(\frac{(k1 + k2)\ x1[t]}{m1}) + \frac{k2\ x2[t]}{m1},$$
$$x2'[t] == v2[t], \quad v2'[t] == \frac{k2\ x1[t]}{m2} - \frac{(k2 + k3)\ x2[t]}{m2}\}$$

DSolve is capable of solving this system symbolically, but it is more instructive to show how *Mathematica* can be used to get the solutions using linear algebra techniques. The first step is to calculate the eigenvalues and eigenvectors of the matrix **A**. We choose some constants at this point to keep the calculations more manageable.

```
springConstants = {k1 -> 1, k2 -> 2, k3 -> 3, m1 -> 4, m2 -> 5};
eigenData = N[ComplexExpand[Eigensystem[A /. springConstants]]]
```

```
{{-1.1573 I, 1.1573 I, -0.640816 I, 0.640816 I},
  {{-0.73307 I, -0.848386, 0.864077 I, 1.},
   {0.73307 I, -0.848386, -0.864077 I, 1.},
   {2.29923 I, 1.47339, 1.56051 I, 1.},
   {-2.29923 I, 1.47339, -1.56051 I, 1.}}}
```

We see that the eigenvalues are purely imaginary. By the general theory, then, we can construct the general solution from the real and imaginary parts of $e^{\lambda t}\mathbf{v}$ where λ is an eigenvalue of **A** with eigenvector **v**. We only need to use two of the eigenvalues as their conjugates will lead to solutions that are linear combinations of solutions already found.

```
twoSpringSoln = (
   c1 Re[Exp[eigenData[[1, 1]] t] * eigenData[[2, 1]]] +
   c2 Im[Exp[eigenData[[1, 1]] t] * eigenData[[2, 1]]] +
   c3 Re[Exp[eigenData[[1, 3]] t] * eigenData[[2, 3]]] +
   c4 Im[Exp[eigenData[[1, 3]] t] * eigenData[[2, 3]]]
   ) //ComplexExpand
```

```
{2.29923 c4 Cos[0.640816 t] - 0.73307 c2 Cos[1.1573 t] +
   2.29923 c3 Sin[0.640816 t] - 0.73307 c1 Sin[1.1573 t],
   1.47339 c3 Cos[0.640816 t] - 0.848386 c1 Cos[1.1573 t] -
   1.47339 c4 Sin[0.640816 t] + 0.848386 c2 Sin[1.1573 t],
   1.56051 c4 Cos[0.640816 t] + 0.864077 c2 Cos[1.1573 t] +
   1.56051 c3 Sin[0.640816 t] + 0.864077 c1 Sin[1.1573 t],
   1. c3 Cos[0.640816 t] + 1. c1 Cos[1.1573 t] -
   1. c4 Sin[0.640816 t] - 1. c2 Sin[1.1573 t]}
```

For particular initial conditions we calculate c_1, c_2, c_3, and c_4 by substituting into our general solution. We consider the solution with initial condition, $x_1(0) = 4$, $v_1(0) = 0$, $x_2(0) = 0$, and $v_2(0) = 0$.

```
constants =
    First[Solve[Thread[(twoSpringSoln /. t -> 0) == {4, 0, 0, 0}],
           {c1, c2, c3, c4}]]

{c1 -> 0., c2 -> -1.99383, c3 -> 0., c4 -> 1.10401}
```

Here is a plot of the velocity of the first mass.

```
Plot[Evaluate[(twoSpringSoln /. constants)[[2]]], {t, 0, 120},
    AxesLabel -> {time, velocity}, AxesOrigin -> {0, -3.5}];
```

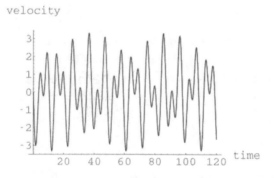

Finally, we can look at the motion of the springs themselves. The printed figure shows only a few frames.

```
spring[a_, b_] := Table[{a + t (b-a), Sin[15 t], Cos[15 t]},
   {t, 0, 1, 0.025}];

Do[ w1 = N[twoSpringSoln[[1]] /. constants];
    w2 = N[twoSpringSoln[[3]] /. constants];
    springPts1 = Join[{{0, 0, 0}, {1, 0, 0}}, spring[1, 7 + w1],
                     {{7 + w1, 0, 0}, {7 + w1 + 3, 0, 0}}];
    springPts2 = Join[{{7 + w1,  0, 0}, {7 + w1 + 3, 0, 0}},
                     spring[10 + w1, 16 + w2],
                     {{16 + w2, 0, 0}, {16 + w2 + 3, 0, 0}}];
    springPts3 = Join[{{16 + w2, 0, 0}, {16 + w2 + 3, 0, 0}},
                     spring[w2 + 19, 25],
                     {{25, 0, 0}, {26, 0, 0}}];
    Show[Graphics3D[{
      {Thickness[0.007], Line[springPts1]},
      {Thickness[0.009], Line[springPts2]},
      {Thickness[0.011], Line[springPts3]},
      Cuboid[{7.5 + w1, 1, 1}, {9.5 + w1, -1, -1}],
      Cuboid[{16.3 + w2, 1.2, 1.2}, {18.7 + w2, -1.2, -1.2}]}],
        PlotRange -> {{0, 26}, {-2, 2}, {-2, 2}},
        PlotLabel -> StringForm["`` seconds", t],
        ViewPoint -> {0, 2, 0}], {t, 0, 49, 0.5}]
```

0 seconds

1. seconds

2. seconds

3. seconds

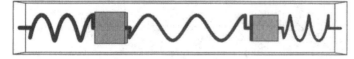

4. seconds

Chapter 9

Logistic Models of Population Growth

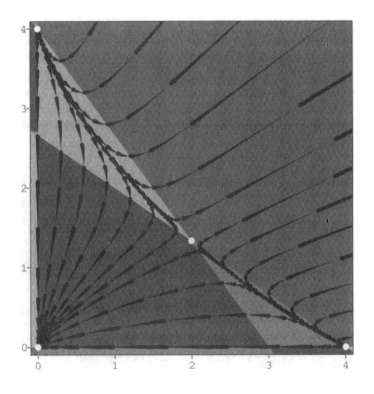

Overview

Differential equations used to model population growth are considered in this chapter. In general it is very difficult to come up with accurate models for even the simplest of ecological systems. In the first section we look at the logistics law of population growth and apply it to the population of the United States. The excellent results, first found by Pearl and Reed [PR], are one of the biggest success stories in the theory of modeling with differential equations. Note that Pearl and Reed did not consider a differential equation but found a population curve based on assumptions of what its characteristics should be. Anyone considering population growth models will find their original paper very interesting.

9.1 ■ One Population

The simplest model of population growth is the exponential growth model, $y' = ay$. One important feature of this model is that y grows without bound over time. Thus the model is only valid for small time intervals over which the population has unlimited space and resources. For longer term predictions we have to take into account the fact that the growth rate will become smaller over time as the population grows to a point where the population will start competing with itself for the available resources.

The simplest way to do this is to assume that the growth rate decreases linearly as the population grows. This is referred to as the logistic law of population growth. This leads to the differential equation: $y' = (a - by)y$ or $y' = ay - by^2$. The growth rate is approximately a for small values of the population and the growth rate becomes small as y approaches a/b; this constant is called the *carrying capacity* of the system. The by^2 term is sometimes thought of as a competition term.

We begin by defining the differential equation for logistics growth in *Mathematica* notation.

```
popEqn = (y'[t] == a y[t] - b y[t]^2);
```

It turns out that we need to load the `Calculus'DSolve'` package to solve the differential equation symbolically with *Mathematica* (this is not necessary in version 3.0). Loading in this package before using `DSolve` is often a good idea.

```
Needs["Calculus'DSolve'"];

popSoln = DSolve[{popEqn, y[t0] == y0}, y[t], t]
```

$$\left\{\left\{y[t] \to \frac{a}{b - E^{-(a\,t)\,+\,Log[(E^{a\,t0}\,(-a\,+\,b\,y0))/y0]}}\right\}\right\}$$

As is the case for exponential growth we can find the coefficients a and b for a particular population if we are given the population at different times. If we know the population at three equally spaced time intervals, then we can find values for a and b. Let's use the times t_0, $t_0 + \delta$, and $t_0 + 2\delta$, with corresponding populations y_0, y_1, and y_2.

```
eqn1 = Simplify[y1 == y[t] /. popSoln[[1]] /. t -> t0 + delta]
```

$$y1 == \frac{a}{b + \frac{a - b\,y0}{E^{a\,delta}\,y0}}$$

```
eqn2 = Simplify[y2 == y[t] /. popSoln[[1]] /.
              {t0 -> t0 + delta, t -> t0 + 2 delta, y0 -> y1}]
```

$$y2 \ == \ \cfrac{a}{b + \cfrac{a - b\ y1}{E^{a\ delta}\ y1}}$$

We would like *Mathematica* to solve the two previous equations for *a* and *b*, but it fails to do so.

```
Solve[{eqn1, eqn2}, {a, b}]
```

$$Solve[\{y1 \ == \ \cfrac{a}{b + \cfrac{a - b\ y0}{E^{a\ delta}\ y0}}, \ y2 \ == \ \cfrac{a}{b + \cfrac{a - b\ y1}{E^{a\ delta}\ y1}}\}, \ \{a, b\}]$$

But we can help things along by observing that both the equations are linear in *b*; thus we first solve each equation for *b*.

```
bFromEqn1 = b /. First[Solve[eqn1, b]]
```

$$-(\cfrac{-(a\ E^{a\ delta}\ y0) + a\ y1}{-(y0\ y1) + E^{a\ delta}\ y0\ y1})$$

```
bFromEqn2 = b /. First[Solve[eqn2, b]]
```

$$-(\cfrac{-(a\ E^{a\ delta}\ y1) + a\ y2}{-(y1\ y2) + E^{a\ delta}\ y1\ y2})$$

We now set the two results equal and solve for *a*.

```
aSoln = a /. First[Solve[bFromEqn1 == bFromEqn2, a]]
```

$$\cfrac{Log[\cfrac{(y0 - y1)\ y2}{y0\ (y1 - y2)}]}{delta}$$

Now we compile a list of the actual population of the U.S. from census reports.

```
actualPop = {{1790, 3929000}, {1800, 5308000}, {1810, 7240000},
    {1820, 9638000}, {1830, 12866000}, {1840, 17069000},
    {1850, 23192000}, {1860, 31443000}, {1870, 38558000},
    {1880, 50156000}, {1890, 62948000}, {1900, 75995000},
    {1910, 91972000}, {1920, 105711000}, {1930, 122775000},
    {1940, 131669000}, {1950, 150697000}, {1960, 179323175},
    {1970, 203302031}, {1980, 226545805}, {1990, 248709873}};
```

Three entries from this table—we'll use 1790, 1850, and 1910—allow us to get values for our model parameters *a* and *b*. Because *b* depends on *a* we use `ReplaceRepeated` (`//.`), which causes the replacement to be done until there are no longer any changes.

```
{aa, bb} = {aSoln, bFromEqn1} //. {a -> aSoln, delta -> 60,
   y0 -> actualPop[[1, 2]], y1 -> actualPop[[7, 2]],
   y2 -> actualPop[[13, 2]]} //N
```

$$\{0.0313395, \ 1.58863 \ 10^{-10}\}$$

With values for *a* and *b* in hand, we can sketch the population predicted by the logistics model (starting from the value for 1790) together with the actual population.

```
Needs["VisualDSolve`"]

popView = VisualDSolve[popEqn /. {a -> aa, b -> bb},
   {t, 1790, 2000}, {y, 0, 2.5 10^8},
   DirectionField -> True,
   InitialValues -> actualPop[[1]],
   PlotStyle -> {AbsoluteThickness[2]},
   WindowShade -> Gray,
   Epilog -> {AbsolutePointSize[5], White, Map[Point, actualPop]}];
```

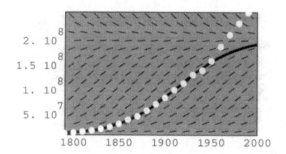

It is clear from the direction field that the model has a limiting population near 200,000,000. It is easy to see that the constant function $y(t) = a/b$ is an equilibrium solution of the equation.

```
{popEqn /. y[t] -> a/b}
```

$$\{y'[t] == 0\}$$

```
aa/bb
```

$$1.97274 \ 10^{8}$$

We see from the diagram that the predicted population is in close agreement with the actual population up to around 1950 after which the predicted population falls far short. In fact, the predicted limit was reached before 1970. This prediction is based on the populations in 1790, 1850, and 1910 and was made by Pearl and Reed in their 1920 paper. The excellent agreement from 1920 to 1950 is one of the more remarkable successes in predicting population growth.

9.2 ▬ Two Populations

In the logistics law of population growth for a single population, $y' = ay - by^2$, the by^2 term is thought of as an internal competition term. This reflects the fact that members of the population compete with themselves for resources, which are limited. For two populations that compete with each other for resources, the simplest logistics-type model we could use would be:

```
pop = {x'[t] == a x[t] - b x[t] y[t] - c x[t]^2,
       y'[t] == d y[t] - e x[t] y[t] - f y[t]^2};
```

Here a, b, c, d, e, and f are positive constants. For most choices of coefficients we will see four equilibrium points:

```
{x[t], y[t]} /. Solve[pop /. {x'[t] -> 0, y'[t] -> 0}, {x[t], y[t]}]
```

$$\left\{\{0, 0\}, \left\{\frac{a}{c}, 0\right\}, \left\{\frac{a}{c} + \frac{b\,(-(c\,d) + a\,e)}{c\,(-(b\,e) + c\,f)}, -\left(\frac{-(c\,d) + a\,e}{-(b\,e) + c\,f}\right)\right\}, \left\{0, \frac{d}{f}\right\}\right\}$$

We first note that if either population is 0, then the other population obeys the logistics law of population growth. To see what happens to the populations when one or the other is 0, we choose some coefficients and use `PhaseLine` to draw the one-dimensional phase portraits.

```
coeffs1 = {a -> 4, b -> 3/2, c -> 2, d -> 3, e -> 1, f -> 3/2};

PhaseLine[pop[[1]] /. y[t] -> 0 /. coeffs1, x[t], {x, 0, 4},
    FlowField -> True, NumberFish -> 25];
```

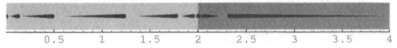

```
    0.5      1      1.5     2     2.5      3      3.5      4
```

```
PhaseLine[pop[[2]] /. x[t] -> 0 /. coeffs1, y[t], {y, 0, 4},
   FlowField -> True, NumberFish -> 25];
```

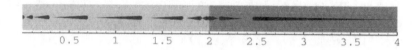

We see that with these coefficients, both populations approach 2. In the case that $y = 0$ this corresponds to c/a; for $x = 0$ it is f/d.

Next, consider the equilibrium points and their eigenvalues corresponding to the linearization about the equilibrium points, for the situation where $x \geq 0$ and $y \geq 0$. A convenient way to do this is to use the built-in capabilities of PhasePlot to calculate equilibrium points and eigenvalues.

```
PhasePlot[pop /. coeffs1, {x[t], y[t]}, {t, -2, 2},
   {x, -0.1, 2.5}, {y, -0.1, 2.5}, AspectRatio -> Automatic,
   ShowEquilibria -> True, ShowEigenvalues -> True];
```

Equilibria   Eigenvalues

{0, 0}   {3, 4}

{1, $\frac{4}{3}$}   {$\frac{-4 - 2\ \text{Sqrt}[2]}{2}$, $\frac{-4 + 2\ \text{Sqrt}[2]}{2}$}

{2, 0}   {-4, 1}

{0, 2}   {-3, 1}

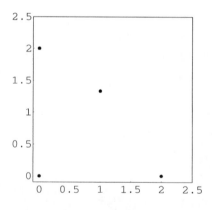

The equilibrium points at (0, 2) and (2, 0) are not stable, because the linearized system has one positive and one negative eigenvalue; but the equilibrium point at (1, 4/3) is stable because both eigenvalues are negative. Using NullclineShading and FlowParametricPlot for some carefully chosen initial values makes it clear what is going on.

```
inits = Join[Table[{0, x, a/b - c/b x}, {x, 0, 4, 0.3}],
             Table[{0, x, d/f - e/f x}, {x, 0, 4, 0.3}],
             {{0, 1, 0}, {0, 0, 1}}];
```

```
PhasePlot[pop /. coeffs1, {x[t], y[t]},
    {t, -2, 2}, {x, -0.1, 4}, {y, -0.1, 4},
    AspectRatio -> Automatic, InitialValues -> (inits /. coeffs1),
    ParametricPlotFunction -> FlowParametricPlot,
    ShowEquilibria -> True,
    EquilibriumPointStyle -> {AbsolutePointSize[5], White},
    StayInWindow -> True, Segments -> 25,
    NumberFish -> 20, FlowThickness -> 0.75,
    NullclinePlotPoints -> 80, NullclineShading -> True];
```

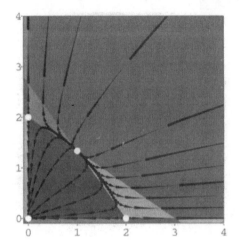

For strictly positive initial values, the populations tend toward the (1, 4/3) equilibrium. An interesting aspect of these equations is the way in which small changes to the environment affect the outcome. If we decrease the internal competition for each species in the model so that c and f have smaller values, then we get a much different outcome. For example, if we cut the values of c and f in half we get the following picture.

```
coeffs2 = {a -> 4, b -> 3/2, c -> 1, d -> 3, e -> 1, f -> 3/4};
```

```
PhasePlot[pop /. coeffs2, {x[t], y[t]},
    {t, -2, 2}, {x, -0.1, 4.1}, {y, -0.1, 4.1},
    AspectRatio -> Automatic,
    InitialValues -> (inits /. coeffs2),
    ParametricPlotFunction -> FlowParametricPlot,
    ShowEquilibria -> True,
    EquilibriumPointStyle -> {AbsolutePointSize[5], White},
    StayInWindow -> True,
    Segments -> 25, NumberFish -> 20,
    FlowThickness -> 0.75,
    NullclineShading -> True, NullclinePlotPoints -> 80];
```

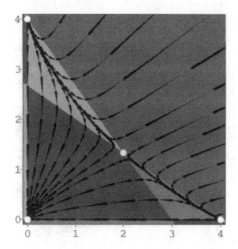

The environment will now support more members of each population, but the end result is that one of the species dies out. If we check the eigenvalues of the linearized systems in this case we will see that (0, 4) and (4, 0) are stable equilibria, but that the equilibrium at (2, 4/3) is unstable.

```
PhasePlot[pop /. coeffs2,  {x[t], y[t]},
   {t, -2, 2}, {x, -0.1, 4.5}, {y, -0.1, 4.5},
   ShowEquilibria -> True, ShowEigenvalues -> True,
   DisplayFunction -> Identity];
```

```
  Equilibria        Eigenvalues
   {0, 0}           {3, 4}
         4             -3 - Sqrt[17]   -3 + Sqrt[17]
   {2,  -}           {-------------, , -------------}
         3                 2               2
   {4, 0}           {-4, -1}
   {0, 4}           {-3, -2}
```

Chapter 10

Hamiltonian Systems

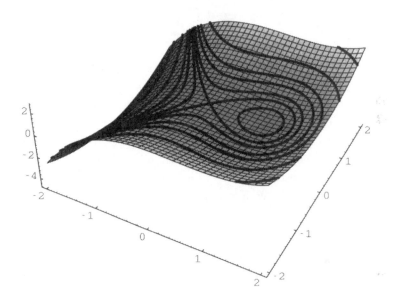

Overview

Hamiltonian systems are special systems of differential equations that arise often in physical models. The existence of a Hamiltonian function can be very helpful in understanding the behavior of solutions to the system.

10.1 ■ An Example

As a first example, consider the system $x' = y$, $y' = x - x^2$. We first look at the flow field, as well as the eigenvalues for the two equilibrium points.

```
ex1 = {x'[t] == y[t], y'[t] == x[t] - x[t]^2};

PhasePlot[ex1, {x[t], y[t]}, {x, -2, 2}, {y, -2, 2},
    FlowField -> True, ShowEquilibria -> True,
    ShowEigenvalues -> True, FieldLogScale -> 6];
```

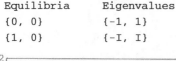

Equilibria	Eigenvalues
{0, 0}	{-1, 1}
{1, 0}	{-I, I}

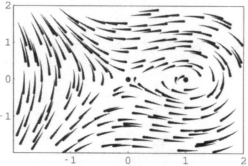

It is not completely clear from the picture whether the orbits circle the point (1, 0), or slowly spiral into it. The eigenvalues for the equilibrium at (1, 0) are i and $-i$, but this sort of information is not definitive. The true behavior might be affected by a small real part that disappears in the linearization (computation of Jacobian). Let's look at some orbits (the initial values have been carefully chosen so that the separatrix— the curve separating the types of solutions—is included; the method for doing this is explained on page 160). We use a larger ComputeWindow so that the orbits that disappear off the right edge reenter the window.

```
PhasePlot[ex1, {x[t], y[t]},
  {t, 0, 10}, {x, -2, 2}, {y, -2, 2},
  ShowEquilibria -> True, MaxSteps -> 1000,
  StayInWindow -> True, ComputeWindow -> {{-2, 5}, {-2, 2}},
  InitialValues -> Join[
    {{-1.430809539360302, 2, {0, 5}},
     {-0.0099669415166751, -0.01}, {0, 5},
     {0.01003361411018955, 0.01, {0, 13}}},
    Table[{x, 0}, {x, -1.8, -0.4, 0.2}],
    Table[{x, 0, {0, 9}}, {x, 0.15, 0.9, 0.2}],
    Table[{0, y}, {y, 0.2, 2, 0.4}]]];
```

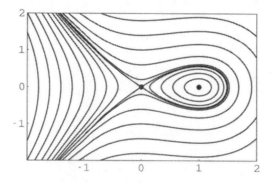

Of course, orbits that are visual loops are not necessarily real loops: perhaps they spiral, but very, very slowly. The fact that this system is Hamiltonian will allow us to say definitively that the loops are loops. A *Hamiltonian function* for a system of two autonomous differential equations ($x' = f$, $y' = g$, where f and g are differentiable functions) is a function $H(x, y)$ such that $H_y = f$ and $H_x = -g$; a *Hamiltonian system* is one that admits a Hamiltonian function. For this particular system, it is not too hard to discover a Hamiltonian function, which we now define as H.

```
H[x_, y_] := y^2/2  - x^2/2 + x^3/3
{D[H[x, y], x], D[H[x, y], y]}
```

$$\{-x + x^2, y\}$$

It is immediate from the definition that a Hamiltonian function does not change along an orbit: just apply the chain rule to show that $d/dt\ H(x(t), y(t)) = 0$ if $(x(t), y(t))$ is a solution to the system. This can be done by *Mathematica* if we first turn the given system into two substitution rules, as follows.

```
rules1 = ex1 /. Equal -> Rule;
D[H[x[t], y[t]], t] /. rules1  //Expand
```

```
0
```

This means that the orbits must lie on the contours of H! We can use Contour-Plot to generate the contours, thus providing independent verification of the geometry of the orbits.

```
cPlot = ContourPlot[H[x, y], {x, -2, 2}, {y, -2, 2},
   ContourShading -> False, PlotPoints -> 70,
   Contours -> {-2, -1, -0.5, -0.3, -0.1, 0, 0.1, 0.3, 0.5, 1, 2}];
```

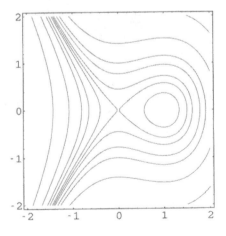

We can illustrate the role *H* plays by looking at the surface that is the graph of *H*. Hamiltonian functions are sometimes called energy functions, the point being that they represent an energy that must remain constant as the orbits evolve. Here is a picture of *H* for our example.

```
surf = Plot3D[H[x, y], {x, -2, 2}, {y, -2, 2},
    ColorFunction -> GrayLevel, PlotPoints -> 40];
```

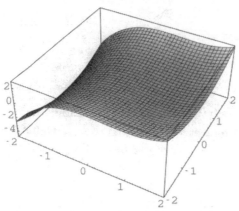

We can superimpose the contours on the surface by using `Graphics[cPlot]` to turn the stored contour plot into a `Graphics` object, using `Cases[..., _Line, Infinity]` to extract all the `Line` objects from this object, using `/.` to turn the pairs of reals in these line objects to points in 3-space lying slightly (0.05) above the surface, using `Graphics3D` to generate some thick lines from this data, and then showing everything together.

```
Show[surf,
  Graphics3D[{AbsoluteThickness[2],
          Cases[Graphics[cPlot], _Line, Infinity] /.
              {x_Real, y_} :> {x, y, H[x, y] + 0.05}}],
    Boxed -> False];
```

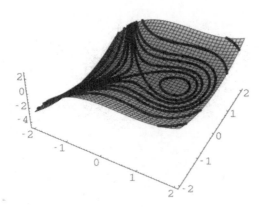

From this point of view it is obvious that the closed loops in our original numerical example are really closed loops. *H* has a local minimum at (1, 0) (easy to verify) and so the contours near that minimum must be loops. Thus the two equilibria really are a saddle and center, as indicated by the eigenvalues and the numerical solution.

Another way to make good use of the Hamiltonian is to check the values of *H* along the numerically computed orbits. Here is how this can be done. First we generate an orbit and save the solution.

```
PhasePlot[ex1, {x[t], y[t]},
    {t, 0, 6}, {x, -2, 2}, {y, -2, 2},
    InitialValues -> {-1.22, 2}, SaveSolution -> True];
```

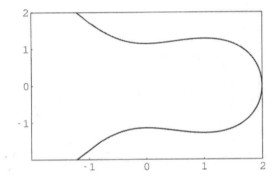

Now we plug the solution (stored in the global variable `Solution`) into *H*, and then substitute 25 *t*-values from the domain. The results are pretty much constant, which shows that the numerical approximation has indeed stayed, within small tolerances, on an orbit of *H*. We use a threading trick to generate the rules. Here's an abstract example.

```
Thread[Rule[{x, y}, {s1, s2}]]

{x -> s1, y -> s2}
```

In the case at hand, `Solution` consists of two expressions in *t*, so we thread over `Solution`.

```
H[x, y] /. Thread[Rule[{x, y}, Solution]] /. t -> Range[0, 6, 0.25]

{0.650517, 0.65052, 0.650521, 0.650521, 0.65052,
    0.65052, 0.650521, 0.650521, 0.650521, 0.650518,
    0.650514, 0.650511, 0.65051, 0.65051, 0.65051,
    0.65051, 0.65051, 0.650511, 0.650513, 0.650511,
    0.650511, 0.650513, 0.650515, 0.650521, 0.65053}
```

10.2 An Ideal Pendulum

An ideal pendulum (having unit length) is governed by the second-order equation $\theta''(t) = -g\sin\theta$. Here we convert that into a system and look at some orbits. The special value of v—$14\sqrt{5}$—is chosen to yield some initial conditions for orbits lying exactly on the separatrices.

```
g = 9.8; v1 = 14/Sqrt[5];
pendulum = {theta'[t] == v[t], v'[t] == - g Sin[theta[t]]};
```

```
PhasePlot[pendulum, {theta[t], v[t]}, {t, 0, 10},
   {theta, -2 Pi, 2 Pi}, {v, -8.5, 8.5},
   ShowEquilibria -> True, DirectionArrow -> True,
   EquilibriumPointStyle -> {AbsolutePointSize[2]},
   InitialValues -> {
     {-2 Pi,    4, { 0, 1.5}}, {-2 Pi,   5.5, { 0, 2  }},
     {    0,    4, { 0, 3  }}, {    0,   5.5, { 0, 5  }},
     { 2 Pi,   -4, { 0, 1.5}}, { 2 Pi, -5.5, { 0, 1.5}},
     {-2 Pi,    8, { 0, 2  }}, { 2 Pi,   -8, { 0, 2  }},
     {-2 Pi,   v1, { 0, 2  }}, {-2 Pi,  -v1, {-2, 0  }},
     { 2 Pi,  -v1, { 0, 2  }}, { 2 Pi,   v1, {-2, 0  }},
     {    0,   v1, {-2, 2  }}, {    0,  -v1, {-2, 2  }}}];
```

```
PhasePlot[pendulum, {theta[t], v[t]}, {t, 0, 10},
   {theta, -2 Pi, 2 Pi}, {v, -8.5, 8.5},
   ShowEquilibria -> True, DirectionArrow -> True,
   NullclineShading -> True,
   FlowField -> True, FieldMeshSize -> 10,
   EquilibriumPointStyle -> {AbsolutePointSize[5], White},
   PlotStyle -> {AbsoluteThickness[1.5], White},
   InitialValues -> {
     {-2 Pi,    4, { 0, 1.5}}, {-2 Pi,   5.5, { 0, 2  }},
     {    0,    4, { 0, 3  }}, {    0,   5.5, { 0, 5  }},
     { 2 Pi,   -4, { 0, 1.5}}, { 2 Pi, -5.5, { 0, 1.5}},
     {-2 Pi,    8, { 0, 2  }}, { 2 Pi,   -8, { 0, 2  }},
     {-2 Pi,   v1, { 0, 2  }}, {-2 Pi,  -v1, {-2, 0  }},
     { 2 Pi,  -v1, { 0, 2  }}, { 2 Pi,   v1, {-2, 0  }},
     {    0,   v1, {-2, 2  }}, {    0,  -v1, {-2, 2  }}}];
```

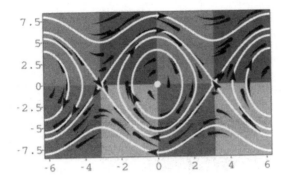

There is a very quick way of telling whether an autonomous system is Hamiltonian. If the system is $x' = f(x, y)$, $y' = g(x, y)$, then the system is Hamiltonian if and only if $f_x + g_y = 0$. (Proof: For the forward direction use the fact that $H_{xy} = H_{yx}$. The reverse direction follows from the discussion to be given shortly on how to use integrals to construct H.) It follows immediately that a decoupled system such as the pendulum ($\theta' = v$, $v' = -g \sin \theta$) is Hamiltonian.

Still, here is how one would use *Mathematica* to check.

```
D[pendulum[[1, 2]], theta[t]] + D[pendulum[[2, 2]], v]
```

```
0
```

Here is a less trivial example.

```
ex2 = {x'[t] == -x[t] Sin[y[t]] + 2 y[t], y'[t] == -Cos[y[t]]};
D[ex2[[1, 2]], x[t]] + D[ex2[[2, 2]], y[t]]
```

```
0
```

Now, in these cases integration methods can be brought to bear to determine a Hamiltonian function. The general method is as follows: Given a Hamiltonian system $x' = f(x, y)$, $y' = g(x, y)$, one first sets H to be the integral of f with respect to y; this gives H as a sum of $s(x, y)$, a single antiderivative, and a varying constant of integration, $C(x)$. Then, differentiating with respect to x, one gets that $C'(x) = -g(x, y) - \partial s / \partial x$. Then $C(x)$ can be found by integration, and we have H.

```
s = Integrate[v, v];
H = s + Integrate[9.8 Sin[theta] - D[s, theta], theta]
```

$$\frac{v^2}{2} - 9.8 \; Cos[theta]$$

Here is how to use the Hamiltonian function to get initial conditions on the separatrix. First compute the energy at the equilibrium.

```
v^2/2 - 98/10 Cos[theta] /. {theta -> -Pi, v -> 0}
```

$$\frac{49}{5}$$

Now, for $\theta = 0$ we can easily find the values of v that yield an energy of $49/5$; this explains our use of the initial condition of $14\sqrt{5}$ to generate the pendulum separatrices.

```
v /. Solve[v^2/2 - 98/10 Cos[theta] == 49/5 /. theta -> 0]
```

$$\{\frac{-14}{\text{Sqrt}[5]}, \frac{14}{\text{Sqrt}[5]}\}$$

We can now automate these procedures to define a Hamiltonian query, and then a routine that produces the Hamiltonian function provided the system is indeed Hamiltonian.

```
HamiltonianQ[{f_, g_}, {x_, y_}] := Simplify[D[f, x] + D[g, y]] == 0

HamiltonianFunction[{f_, g_}, {x_, y_}] :=
  Module[{s = Integrate[f, y]},
     s - Integrate[g + D[s, x], x]] /; HamiltonianQ[{f, g}, {x, y}]
```

For the two examples of this section our algorithm has no difficulty in finding a Hamiltonian function.

```
HamiltonianFunction[{v, -9.8 Sin[theta]}, {theta, v}]
```

$$\frac{v^2}{2} - 9.8 \text{ Cos[theta]}$$

```
HamiltonianFunction[{-x Sin[y] + 2 y, -Cos[y]}, {x, y}]
```

$$y^2 + x \text{ Cos[y]}$$

10.3 ■ Higher Dimensions

Hamiltonians can be useful for larger systems, such as the system describing the motion of one planet around a sun. Such motion is described by the following system (see page 108).

```
planets = {x'[t] == u[t],
           y'[t] == v[t],
           u'[t] == -x[t] / (x[t]^2 + y[t]^2)^(3/2),
           v'[t] == -y[t] / (x[t]^2 + y[t]^2)^(3/2)};
```

A function $H(x, y, u, v)$ is a Hamiltonian function for such a system if $H_x = -u'$, $H_y = -v'$, $H_u = x'$, and $H_v = y'$. It is easy to check by the chain rule that such an H must be conserved along orbits. For planets, the following is a Hamiltonian function (it is a sum of kinetic energy and potential energy).

```
Clear[H]
H = 1/2 (u^2 + v^2) - 1/Sqrt[x^2 + y^2];
{D[H, x], D[H, y], D[H, u], D[H, v]}
```

$$\left\{ \frac{x}{(x^2 + y^2)^{3/2}}, \frac{y}{(x^2 + y^2)^{3/2}}, u, v \right\}$$

As in the two-dimensional case, we can check the values of H along some computed orbits, which is a good check on the accuracy of the numerical solution. Because turning SaveSolution on causes functions for only the plotted variables to be saved, we use SecondOrderPlot to plot all four functions, x, y, x', and y'.

```
SecondOrderPlot[{x"[t] == -x[t] / (x[t]^2 + y[t]^2)^(3/2),
                 y"[t] == -y[t] / (x[t]^2 + y[t]^2)^(3/2)},
    {x[t], y[t]}, {t, 0, 2 Pi},
    InitialValues -> {1/2, 0, 0, Sqrt[3]},
    SaveSolution -> True];
```

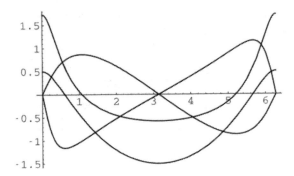

```
H /. Thread[Rule[{x, y, u, v}, Solution]] /.
      t -> Range[0, 2 Pi, 1/4]
```

```
{-0.5, -0.500008, -0.500005, -0.500005, -0.500005,
  -0.500005, -0.500005, -0.500004, -0.500004, -0.500004,
  -0.500004, -0.500004, -0.500004, -0.500004, -0.500004,
  -0.500004, -0.500004, -0.500004, -0.500005, -0.500006,
  -0.500008, -0.50001, -0.500013, -0.500013, -0.500005,
  -0.500003}
```

As before, the numerical algorithm performed admirably, and the energy is near-constant. This method is more useful on sensitive or chaotic systems, such as a double pendulum (see Chapter 14).

Chapter 11

A Devilish Equation

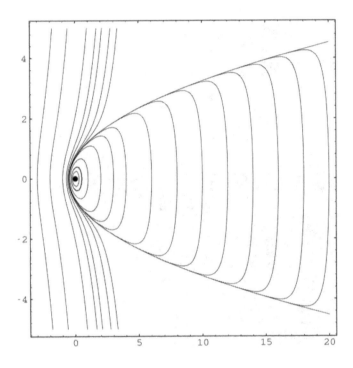

Overview

Sometimes equations that look simple cause tremendous difficulties for numerical algorithms. This chapter contains an in-depth discussion of one such equation. Users of software must always be aware of the need to look critically at the output in the hope of catching those cases, not at all rare, where numerical instabilities cause the algorithms to return incorrect images.

11.1 ░ Skepticism Rewarded

John Polking (Rice University) has found an equation whose simplicity belies its nasty behavior. It's the second-order equation $x'' = (x')^2 - x$, but we may as well look at the corresponding first-order system. Although it is hardly necessary, we may as well use the `VisualDSolve` function `ToSystem` to do the transformation for us.

```
Needs["VisualDSolve'"]
(Polking = ToSystem[x"[t] == x'[t]^2 - x[t], x[t], t,
  AuxiliaryVariables -> y]) //TableForm
```

x'[t] == y[t]

y'[t] == -x[t] + y[t]^2

Now we can start by looking at a vector field, nullclines, and equilibria.

```
PhasePlot[Polking, {x[t], y[t]},
  {t, 0, 5}, {x, -2, 2}, {y, -5, 5},
  Nullclines -> True, NullclineShading -> True,
  ShowEquilibria -> True, ShowEigenvalues -> True,
  VectorField -> True, FieldMeshSize -> 25,
  FieldLogScale -> 5];
```

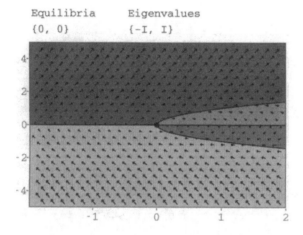

```
Equilibria      Eigenvalues
 {0, 0}          {-I, I}
```

The eigenvalue information in this case does not determine whether the equilibrium is stable. Further analysis will show that all nearby orbits are periodic, and thus it is stable. The parabolic nullcline is just the graph of $x = y^2$. We can look a little more closely inside the parabola; we may as well insert the equilibrium point as an Epilog, to save computation time.

```
equiPt = {AbsolutePointSize[4], Point[{0, 0}]};
```

```
PhasePlot[Polking, {x[t], y[t]},
   {t, 0, 5}, {x, -0.2, 2}, {y, -1.4, 1.4},
   Nullclines -> True, ShowEquilibria -> True,
   VectorField -> True, Epilog -> equiPt];
```

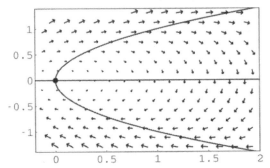

This gives us some idea of the orbit geometry. Now let us look at the orbit starting at (14, 0).

```
PhasePlot[Polking, {x[t], y[t]},
   {t, 0, 16.5}, {x, -2, 20}, {y, -8, 8},
   Epilog -> equiPt, MaxSteps -> 1000, InitialValues -> {14, 0}];
```

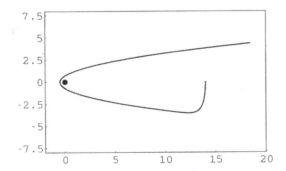

Although not immediately evident, this solution is faulty. In fact, it is not hard to see that if the true orbit starting from (14, 0) really crosses the *x*-axis again as shown, then it must form a closed loop. This follows from the symmetry of the underlying equation: the slopes above the *x*-axis mirror the ones below, so if we work backwards in time from (14, 0) we must strike the identical second-crossing.

In fact, when we look at this backward orbit, it becomes immediately clear that the computed solution violates the uniqueness theorem, since it gives an orbit that crosses itself! Here is how to look both forward and backward. We use `PhasePlot`, and its ability to accept varied *t*-domains, so that we can shade the forward and backward parts differently. Careful examination will show that a crossing exists.

```
PhasePlot[Polking, {x[t], y[t]},
   {t, 0, 16.5}, {x, -2, 20}, {y, -8, 8},
   Epilog -> equiPt, MaxSteps -> 1000,
   ShowInitialValues -> True,
   PlotStyle -> {{Gray, AbsoluteThickness[4]}, {}},
   InitialValues -> {{14, 0}, {14, 0, {-16.5, 0}}}];
```

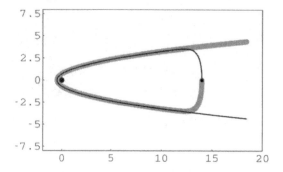

Let's zoom in and look at some orbits near the equilibrium point to improve our understanding.

```
PhasePlot[Polking, {x[t], y[t]}, {t, 0, 4}, {x, -2, 2}, {y, -5, 5},
   Epilog -> equiPt, StayInWindow -> True,
   InitialValues -> {{-2, 0}, {-1.5, 0}, {-1, 0}, {-0.8, 0}, {-0.6, 0},
      {-0.55, 0}, {-0.45, 0}, {-0.4, 0}, {-0.2, 0}, {-0.1, 0}}];
```

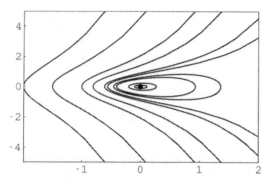

This image seems to make the orbit pattern pretty clear: it looks as if there are two types of solutions, bounded loops and doubly infinite curves. We will show in section 11.2 that this is indeed the case. There is a curve that separates the bounded orbits from the unbounded ones; such a curve, called a *separatrix*, is itself an orbit and, as we shall see, it is simply a parabola in this case.

But closer inspection shows that something nasty is happening near $(-0.5, 0)$: it looks like a tremendous amount of compression is taking place. If the numerical approximation is only a little bit off in this region, it can easily get swept away on a very different orbit.

We convinced ourselves above that something went wrong for (14, 0). Another approach to detecting such errors is to redo the computation with more working precision and less local truncation error. This is achieved by increasing the `AccuracyGoal` and `PrecisionGoal` options, and increasing the `MaxSteps` option because of the extra work that will be required.

```
PhasePlot[Polking, {x[t], y[t]}, {t, 0, 18}, {x, -2, 20}, {y, -8, 8},
    Epilog -> equiPt, InitialValues -> {14, 0}, MaxSteps -> 5000,
    AccuracyGoal -> 15, PrecisionGoal -> 15, WorkingPrecision -> 20];
```

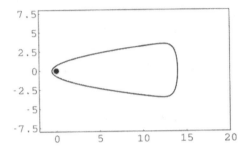

This illustrates the general principle of numerical approximation: If an increase in working precision leads to different results at low precision, then something is definitely wrong with the low-precision result. Often such behavior is evidence that the high-precision computation is wrong too, but in this case the looping behavior is indeed correct.

A curiosity of this equation is that, for this particular initial value, the standard (nonadaptive) fourth-order Runge–Kutta method outperforms the default adaptive routine in *Mathematica.* Here we use 2000 Runge–Kutta steps, and the closed loop shows up! We leave it to the reader to investigate this further and perhaps provide an explanation. Of course, as we will see in section 11.2, the difficulty cannot be put off indefinitely; any method will have great trouble as the initial value moves out the *x*-axis.

```
PhasePlot[Polking, {x[t], y[t]}, {t, 0, 18},
    {x, -2, 20}, {y, -8, 8}, Method -> RungeKutta4,
    RKSteps -> 2000, Epilog -> equiPt, InitialValues -> {14, 0}];
```

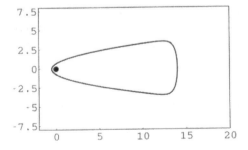

11.2 | **What's Going On**

We now bring some qualitative methods to bear on this equation. The analysis here is due to John Polking; of course, in general such a detailed analysis of a computationally nasty problem will be impossible. Indeed, it can be quite a challenge simply to recognize that a problem has the potential for bad behavior.

The symmetry of the equation implies that for a solution $(x(t), y(t))$, x, when viewed (locally) as a function of y, is an even function. Thus we can write $x = f(y^2)$. The inverse of this relationship yields $y(t)^2 = g(x(t))$. Differentiate with respect to t to get $2y(t)y'(t) = g'(x(t))x'(t)$.

Using the original differential equation we can simplify the previous equation to: $2(y^2 - x) = g'(x)$, or $g'(x) - 2g(x) = -2x$. We can solve this new differential equation, now including an initial condition on the x-axis: $g(x_0) = 0$.

```
g[x] /. First[
   DSolve[{g'[x] - 2 g[x] == - 2 x, g[x0] == 0}, g[x], x]]
```

$$\frac{1}{2} + x - \frac{E^{2\ x\ -\ 2\ x0}\ (1\ +\ 2\ x0)}{2}$$

Since this last expression is g, which equals y^2, we can conclude that any orbit starting on the x-axis satisfies

$$(2x_0 + 1)e^{2(x - x_0)} - 2x - 1 + 2y^2 = 0$$

When $x_0 = -1/2$, we get the special parabolic solution curve $y^2 - x - 1/2 = 0$, and this is in fact the separatrix referred to earlier. This particular solution is given explicitly by $x = t^2/4 - 1/2$, $y = t/2$, as can be worked out by hand or by computer, as follows:

```
DSolve[Join[Polking, {x[0] == -1/2, y[0] == 0}] /.
       x[t] -> y[t]^2 - 1/2, {x[t], y[t]}, t]
```

$$\{\{x[t] \rightarrow -(\frac{1}{2}) + \frac{t^2}{4},\ y[t] \rightarrow \frac{t}{2}\}\}$$

The parabolic solution separates the plane into two regions in which the solution curves have different features. To the left of the parabola, where $x_0 < -1/2$, the solution curves are unbounded. To the right, where $x_0 > -1/2$, the solution curves are closed. And the origin is, as we have seen, an equilibrium point. The phase plane image now illustrates the computational difficulties. Given any positive x_0, no matter how large, the solution curve must intersect the x-axis in another point x_1, with $-1/2 < x_1 < 0$. And we can say quite a bit about x_1 because it satisfies

$$(2x_0 + 1)e^{2(x_1 - x_0)} - 2x_1 - 1 = 0$$

which reduces to

$$2x_1 + 1 = (2x_0 + 1)e^{2(x_1 - x_0)}.$$

But the left side is twice the distance from $-1/2$ to x_1. So this distance is asymptotic to

$$(x_0 + 1/2)e^{-2(x_1 - x_0)}$$

Since $x_1 < 0$, this is at most $(x_2 + 1/2)e^{-2x_0}$, which is of course very small for modest values of x_0. When $x_0 = 14$, this is about 10^{-11}. When $x_0 = 20$, the distance is at most 10^{-17}, which is less than machine precision. So there is essentially no hope of coming out on the right orbit.

We may as well use the algebraic formulation to dodge all these numerical issues entirely and generate a nice image of the orbits. We do this by using a contour plot to define `PolkingOrbit[x0]` to generate a single solution starting at $(x_0, 0)$, and then showing them all at once. Alternatively, one could actually solve for y as a function of x and plot the resulting curves.

```
PolkingOrbit[x0_] :=
  ContourPlot[(2 x0 + 1) Exp[2(x - x0)] - 2x - 1 + 2y^2,
    {x, -3, 20}, {y, -5, 5}, Contours -> {0},
    ContourShading -> False, PlotPoints -> 120,
    DisplayFunction -> Identity]

Show[Map[PolkingOrbit, {-3, -2, -1, -0.7, -0.6, -0.53,
    0.52, -0.51, -0.5, 0.3, 0.5, 1, 2, 3, 4, 6, 8, 10, 12,
    14, 16, 18, 20}],
    DisplayFunction -> $DisplayFunction, Epilog -> equiPt];
```

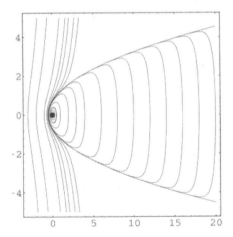

The main problem with this example is the tremendous convergence, followed by rapid divergence. This is quite common (for more examples see sections 1.6 and 15.2); clearly the user of numerical software for differential equations must be constantly aware of the potential for difficulties caused by this, and other problems that can foil even the best numerical algorithms.

Chapter 12

Lead Flow in the Human Body

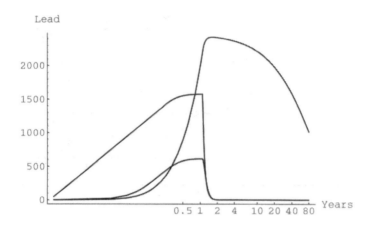

Overview

We discuss here a typical compartment model, in which the effect of a chemical (lead) on a human is analyzed by isolating the effects on different compartments (blood, bones, tissues). This discussion is adapted from the presentation in [BCB, pp. 223–226], which in turn is based on experimental investigations reported in [BBS] and [RWK].

12.1 ■ The Model

When living in a heavily populated and industrial environment, a human becomes contaminated by lead that enters the body through food, air, and water. The lead enters the bloodstream first, and then accumulates in the blood, tissues, and, especially, the bones. Although the body eliminates some of the lead through urine and the loss of hair, nails, and sweat, enough may remain to cause serious health problems.

To model the flow of lead over time, we divide the body into three compartments: (1) blood, (2) tissue, and (3) bone. The model is based on the following flow chart.

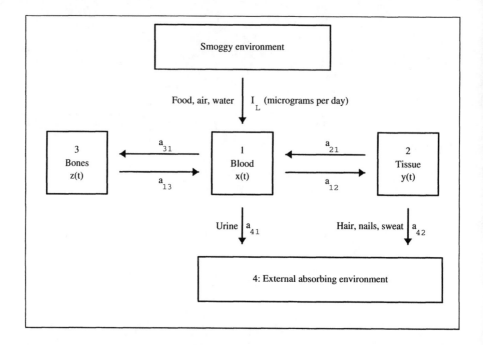

The amount of lead in compartments 1, 2, and 3 at time t is denoted by $x(t)$, $y(t)$, $z(t)$, respectively. The rate of transfer of lead from compartment j to compartment i at time t follows a first-order rate law: this means that, as with so many other differential equation modeling problems, the rate of transfer out of a compartment is just proportional to how much is in the compartment being exited. The constant for flow into compartment i is denoted a_{ij}. Note that $a_{ij} \geq 0$ and that $a_{ij} = 0$ if no transfer takes place. The constant for the reverse flow (from i to j, denoted a_{ji}) is not generally equal to a_{ij}. In other words, the amount of lead flowing from blood to bones is proportional to how much is in the blood; the amount flowing from bones to blood is proportional to how much is in the bones); these rate constants will not be the same.

Further, let I_L (LeadIn in the code) be the net input flow into the bloodstream, let a_{41} be the rate constant for lead loss in urine, and let a_{42} be the rate constant for lead loss from the tissues (hair, nails, sweat). Thus we consider compartment 4 to be the external absorbing environment.

Here are some rate constants, determined experimentally by studying a healthy volunteer living in an area of heavy smog (Los Angeles). These constants were determined by Rabinowitz *et al* ([RWK]; see also [BBS]) and it must be emphasized that they apply only to the one individual whom they studied in detail.

$$a_{12} = 0.0124 \qquad a_{21} = 0.0111$$
$$a_{31} = 0.0039 \qquad a_{13} = 0.000035$$
$$a_{41} = 0.0211 \qquad a_{42} = 0.0162$$

And I_L was observed to be 49.3 micrograms per day (this is not a rate constant; this is the absolute amount coming in).

Now, each compartment obeys a balance law:

The net rate of accumulation of lead in the compartment = Rate in − Rate out

This leads to the following system of differential equations, where t is measured in days.

```
a12 = 0.0124;   a21 = 0.0111;   a31 = 0.0039;   a13 = 0.000035;
a41 = 0.0211;   a42 = 0.0162;   LeadIn  = 49.3;

(LeadFlowEqns = {
   x'[t] == -(a41 + a21 + a31) x[t] + a12 y[t] + a13 z[t] + LeadIn,
   y'[t] ==    a21 x[t] - (a42 + a12) y[t],
   z'[t] ==    a31 x[t] - a13 z[t]})    //TableForm

x'[t] == 49.3 - 0.0361 x[t] + 0.0124 y[t] + 0.000035 z[t]
y'[t] == 0.0111 x[t] - 0.0286 y[t]
z'[t] == 0.0039 x[t] - 0.000035 z[t]
```

Now we can quickly get a view of the situation after 400 days, assuming our subject starts with a completely lead-free body. We use a thin line for x, a medium line for y, and a thick line for z.

```
Needs["VisualDSolve`"];

SystemSolutionPlot[LeadFlowEqns, {x[t], y[t], z[t]}, {t, 0, 400},
   InitialValues -> {0, 0, 0}, SolutionName -> "Lead",
   PlotStyle -> {{AbsoluteThickness[0.5]},
       {AbsoluteThickness[1.5]}, {AbsoluteThickness[3]}}];
```

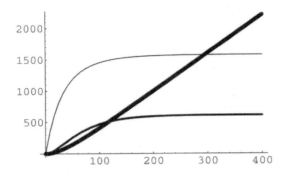

Because we stored the solution in Lead, we can see the levels after 400 days.

```
Day400Values = Lead /. t -> 400
{1577.66, 611.957, 2215.89}
```

We store the largest value for later use.

```
BoneLevel = Last[Day400Values];
```

12.2 ■ Getting the Lead Out

A first question suggested by the graphs is whether the lead in the body will settle down to an equilibrium. The Equilibria function only works on systems of two equations, so we must use a direct approach to find the equilibrium values. We just solve the appropriate linear system.

```
coeffs =  {{ -(a41 + a21 + a31),      a12,       a13 },
          {          a21,       -(a42 + a12),    0  },
          {          a31,            0,      -a13 }};
b = {-LeadIn, 0, 0};
eqVals = LinearSolve[coeffs, b]

{1800.1, 698.639, 200582.}
```

This indicates that the subject's blood will contain 1800 micrograms of lead before equilibrium is reached. The United States Occupational Safety and Health Administration requires that a worker with more than 400 micrograms of lead per liter of blood be removed from the workroom exposed to lead. For a typical person (5 liters of blood) that would be a total blood-lead content of 2000 micrograms, not much above the equilibrium value of 1800. It must be noted that this data comes from a 1973 study; at that time leaded gas was in wide use and no doubt contributed to high levels in humans.

In fact, the system of three differential equations is just linear and a closed-form solution exists; we can find it using DSolve.

```
DSolve[Join[LeadFlowEqns, {x[0] == 0, y[0] == 0, z[0] == 0}],
   {x[t], y[t], z[t]}, t]
```

$$\left\{\left\{x[t] \rightarrow 1800.1 - \frac{719.885}{E^{0.0446688\,t}} - \frac{855.314}{E^{0.0200356\,t}} - \frac{224.898}{E^{0.0000306322\,t}},\right.\right.$$

$$y[t] \rightarrow 698.639 + \frac{497.283}{E^{0.0446688\,t}} - \frac{1108.54}{E^{0.0200356\,t}} -$$

$$\frac{87.3791}{E^{0.0000306322\,t}}, \quad z[t] \rightarrow$$

$$\left.\left.200582. + \frac{62.902}{E^{0.0446688\,t}} + \frac{166.781}{E^{0.0200356\,t}} - \frac{200812.}{E^{0.0000306322\,t}}\right\}\right\}$$

The equilibrium values resulting from the closed form agree with the earlier computation. We can now determine how close we are to equilibrium after 400 days.

```
Day400Values / eqVals
```

```
{0.876428, 0.875928, 0.0110473}
```

Thus the 400-day lead concentrations are within 88%, 88%, and 1% of the equilibrium levels for x, y, and z, respectively. Even after 100 years, the lead in the bones will be only at 67% of equilibrium; thus, in the present scenario, lead will increasingly accumulate in the bones throughout a human's life.

We can now examine the effects of various factors on the levels of lead. For example, suppose that after 400 days the subject moves to northern Minnesota, a totally unleaded environment. How long will it take before all three lead-levels have dropped back to approximately half their 400-day values? Although one could attack this by working out closed forms using cases, it is easier to stick to a numerical approach. We can just use an `If` statement in the equation. The following plot goes out to 80 years.

```
SystemSolutionPlot[{
    x'[t] == -(a41 + a21 + a31) x[t] + a12 y[t] + a13 z[t] +
                If[t < 400, LeadIn, 0],
    y'[t] == a21 x[t] - (a42 + a12) y[t],
    z'[t] == a31 x[t] - a13 z[t]}, {x[t], y[t], z[t]}, {t, 0, 80*365},
    SolutionName -> "LeadAfterMove",
    InitialValues -> {0, 0, 0}, AxesOrigin -> {0, 2500}];
```

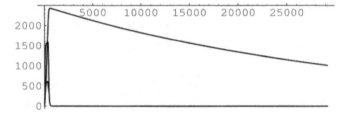

The scale makes the image hard to read. A logarithmic time scale helps; we can get that by loading the `Graphics`Graphics`` package, which contains several logarithmic plotting utilities. And we use the `Ticks` option to customize the tick marks.

```
Needs["Graphics`Graphics`"];
```

```
LogLinearPlot[Evaluate[LeadAfterMove], {t, 1, 365 * 80},
    Ticks ->
        {Map[{365 #, #} &, {0.5, 1, 2, 4, 10, 20, 40, 80}],
        Automatic},
    AxesLabel -> {"Years", "Lead"},
    PlotPoints -> 100];
```

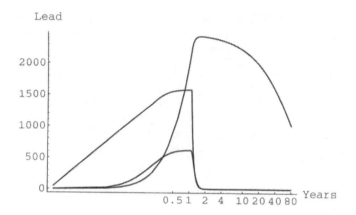

The output shows that x and y drop to 0 within a few months, but z, the bone level, does not. Because we saved the solution in `LeadAfterMove`, we can use `FindRoot` to determine exactly how long, after the 400-day exposure, it takes for the bone level to be cut in half.

```
FindRoot[LeadAfterMove[[3]] == BoneLevel/2, {t, 25000}]

{t -> 26246.9}


(t - 400) / 365 /. %

70.8134
```

Thus it would take over 70 years for the lead in the bones to return to half the level it reached after 400 days.

Certain drugs alleviate the effects of lead poisoning by increasing the rate of removal of lead from the bones back into the blood. The drug-free rate constant for that removal, a_{13}, was given above as 0.000035. Suppose that after 400 days the person both moves to an unleaded environment and starts taking a drug that multiplies that constant by 10, to 0.00035. Under this scenario, the amount of time required to reduce the bone-lead level to half its 400-day level is substantially reduced, but it is still over 7 years.

```
SystemSolutionPlot[{
  x'[t] == -(a41 + a21 + a31) x[t] + a12 y[t] +
                 If[t < 400, a13 z[t] + LeadIn, 10 a13 z[t]],
  y'[t] == a21 x[t] - (a42 + a12) y[t],
  z'[t] == a31 x[t] -   If[t < 400, 1, 10] a13 z[t]},
  {x[t], y[t], z[t]}, {t, 0, 3650},
  SolutionName -> "AfterMoveAndDrugs", InitialValues -> {0, 0, 0},
  PlotRange -> All];
```

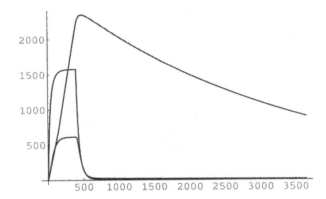

We proceed as before to find the half-life for $z(t)$, the level of lead in the bones.

```
(t - 400) / 365 /.
   FindRoot[AfterMoveAndDrugs[[3]] == BoneLevel/2, {t, 2000}]

7.08835
```

Finally, we address the following question put forward in [BCB]. Suppose one wishes to cut the amount of lead in the bones in half within one year after both starting a drug treatment and moving to a lead-free environment, assuming that 400 days were spent accumulating lead according to the protocol above. What would the efficiency of the drug have to be? In other words, what would the drug have to do to the constant a_{13} in order for the goal to be met?

We solve this by trial and error; that is, we simply try various values of the drug constant k. We can be efficient by not displaying the solution, but rather just storing it and printing out the reduction factor paired with the drug constant.

```
Do[SystemSolutionPlot[{
    x'[t] == -(a41 + a21 + a31) x[t] + a12 y[t] +
                If[t < 400, a13 z[t] + LeadIn, k a13 z[t]],
    y'[t] == a21 x[t] - (a42 + a12) y[t],
    z'[t] == a31 x[t] - If[t < 400, a13, k a13] z[t]},
    {x[t], y[t], z[t]}, {t, 0, 765},
    InitialValues -> {0, 0, 0}, SaveSolution -> True,
    DisplayFunction -> Identity];
  Print[{k, Solution[[3]] / BoneLevel /. t -> 765}],
 {k, 30, 100, 10}]

{30, 0.789242}
{40, 0.706256}
{50, 0.632215}
{60, 0.56614}
{70, 0.507157}
{80, 0.454493}
{90, 0.407456}
{100, 0.365434}
```

Thus we see that, for the specific goals of this example, the drug would have to improve the bone-to-blood constant by a factor of about 70. For a more exact answer, use FindRoot as follows.

```
reduction[k_] := (
  SystemSolutionPlot[{
    x'[t] == -(a41 + a21 + a31) x[t] + a12 y[t] +
              If[t < 400, a13 z[t] + LeadIn, k a13 z[t]],
    y'[t] == a21 x[t] - (a42 + a12) y[t],
    z'[t] == a31 x[t] - If[t < 400, a13, k a13] z[t]},
    {x[t], y[t], z[t]}, {t, 0, 765},
    InitialValues -> {0, 0, 0}, SaveSolution -> True,
    DisplayFunction -> Identity];
  Solution[[3]] / BoneLevel /. t -> 765)

k /. First[FindRoot[reduction[k] == 0.5, {k, {30, 100}}]]
71.2944
```

Chapter 13

Making a Discus Fly

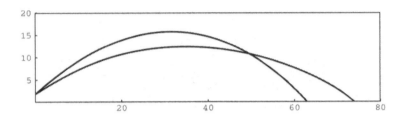

Overview

Discus throwers have long known that a properly thrown discus goes farther against a headwind than with a following wind. The point is that a discus does in fact fly and, under certain wind conditions, the amount of lift it generates can be substantial. Our goal here is to set up a plausible model and see how large the difference in the length of a throw might be. An extra twist is the need to figure out what it means to throw a discus properly in various wind conditions. The details of the analysis presented here are taken from [Fro], a good source for further discussion of this model, and how it might be modified to take more aspects of physical reality into account.

13.1 ■ The Model

Before diving into the details of discus flight, the reader should review the much simpler projectile-with-drag example presented on page 109. Much more information on the analysis of projectiles in sports can be found in [deM].

The key aspect in the modeling of discus flight is an understanding of the coefficients of lift and drag. These depend on the angle of attack: a discus standing on its edge in a horizontal wind creates a much larger drag force than one that is slicing through the air horizontally. The angle of attack is the angle that the discus is inclined to the flow of air. However, this and related angles are a little tricky to compute. The following figure shows the situation.

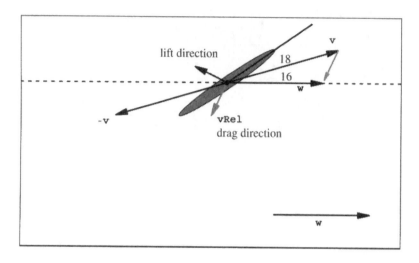

Figure 13.1: The many vectors associated with a flying discus.

The figure shows the situation for a discus flying through the air with a tailwind (represented by **w**, which is assumed to be purely horizontal). The thrower controls two important angles. First, he or she will release the discus in a certain orientation to the ground; we call this angle, the one between the plane of the discus and the ground, the inclination (denoted by α), so as to avoid confusion with the angle of attack, which, as we will see, is not the same thing. We assume that, as is definitely not the case for a football or javelin, α does not change during flight. In the figure α is 34°. The second angle controlled by the thrower is the initial angle of the discus' trajectory, which we call the release angle. Of course, this angle does change during the flight: at the maximum height this angle, the angle of motion, is 0°. In Figure 13.1 this angle is represented by the velocity vector **v**, and is 16°.

The motion of the discus as just described means that the actual passage of wind over the discus occurs not in the direction of the prevailing wind, but in the direction of a vector obtained by combining the wind's vector with the discus's velocity vector. Call this vector $\mathbf{v}_{rel}$; it is just the vector sum of $-\mathbf{v}$ and **w** and is shown in gray in Figure 13.1. Let β denote the angle made by $\mathbf{v}_{rel}$ with the horizontal, about $-117°$ in the example; β is given by $\arctan[y'(t)/x'(t) - w]$. Now comes the key step: the angle of attack—the angle at which the discus is inclined to the air flow (for which we use θ)—is given by $|\beta - \alpha|$. We leave the verification of the several cases to the reader. In Figure 13.1, θ works out to 151°, which fits the picture, since the discus is sharply laid back with respect to the flow of air in the direction of $\mathbf{v}_{rel}$.

Finally, we must understand the directions in which the coefficients of drag and lift have their effect. The drag acts in the direction of airflow, and so in the direction of $\mathbf{v}_{rel}$; the lift acts in the perpendicular direction.

We use C_d and C_l for the coefficients of lift and drag, respectively (explicit values are in the next section), and we assume that the drag and lift forces are proportional to the square of the air speed. The angle of the wind over the discus, β, changes in time, so we use $\beta(t)$. And the angle of attack, θ, is also a function of t, and it is, as noted, just $|\beta(t) - \alpha|$. Let g and k, respectively, denote gravitational acceleration and the constant $\rho A/(2M)$, where A is the cross-sectional area of the discus, M is its mass, an ρ is the density of air; note that $\mathbf{v}_{rel}$ is a function of t too; the speed through the air is $|\mathbf{v}_{rel}(t)|$. The use of the constant k, in particular the fact that $1/2|\mathbf{v}_{rel}(t)|^2$ is used, is explained in [Fro]; these details are fairly standard in the area of fluid flow, and justifications can be found in any fluid flow textbook. Now, a standard application of Newton's law of motion, together with the decomposition of the lift and drag vectors into their x- and y-components, yields the following second-order system that governs the discus's flight. Note that θ is used to get the drag and lift coefficients, but β is used to determine how the coefficients get used.

$$x''(t) = k\left(C_d(\theta(t))(-\cos\beta(t)) + C_l(\theta(t))(-\sin\beta(t))\right)|\mathbf{v}_{rel}(t)|^2$$
$$y''(t) = k\left(C_d(\theta(t))(-\sin\beta(t)) + C_l(\theta(t))(\cos\beta(t))\right)|\mathbf{v}_{rel}(t)|^2 - g$$

13.2 ■ Drag and Lift

To make further progress we need drag and lift data. Using a few points from wind-tunnel experiments, we can set up piecewise linear approximations to the drag and lift that the model can then use as functions. The drag and lift data come from actual wind tunnel experiments (summarized in [Fro]). We work in degrees throughout!

We set these up as functions because they will be used that way throughout the process of solving the differential equation. This is because the angle of flight will change, and so the angle of the air flowing over the discus will be changing during the flight. We use the case restrictor /; (to be read as "If") to define the various cases.

```
Cd[theta_] :=                    0.06        /;   0 <= theta <=  5
Cd[theta_] := 0.0192 theta  - 0.036          /;   5 <= theta <= 30
Cd[theta_] := 0.0115 theta  + 0.195          /;  30 <= theta <= 70
Cd[theta_] := 0.0035 theta  + 0.755          /;  70 <= theta <= 90
Cd[theta_] := Cd[180 - theta]                /;  90 <= theta

Cl[theta_] :=    0.03125 theta               /;   0 <= theta <= 28
Cl[theta_] := -0.0393   theta + 1.9755       /;  28 <= theta <= 35
Cl[theta_] := -0.00714  theta + 0.8498       /;  35 <= theta <= 70
Cl[theta_] := -0.0175   theta + 1.575        /;  70 <= theta <= 90
Cl[theta_] := -Cl[180 - theta]               /;  90 <= theta
```

```
Plot[{Cd[theta], Cl[theta]}, {theta, 0, 180},
   Ticks -> {Range[30, 180, 30], Automatic},
   PlotStyle -> {{Thickness[0.009]}, {}}];
```

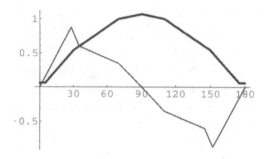

The thicker curve in this plot represents the drag. Note that an angle of attack of 90° maximizes the drag, as it should, and yields zero lift. And an angle between 90° and 180° causes negative lift, because the discus is not "attacking" the air flow.

13.3 ■ Implementing the Equations

An efficient way to set up the system is to keep the equations that SecondOrderPlot sees as simple as possible, putting the somewhat messy algebra, trigonometry, and initializations into a separate routine. The Initialize routine that follows serves that purpose and allows us to control the wind speed and the angles of release and inclination. The height of release is 1.8 meters above the ground, the initial release velocity is 25 meters per second, degree is a conversion factor, and so on (area is in square meters, mass in kilograms, and air density in kg per cubic meter). Observe that each of beta, theta, drag, lift, and vRel2 involves x'[t] and y'[t]. When they occur in the differential equations (DiscusEqns) they will be properly interpreted in terms of the ever-changing $x'(t)$ and $y'(t)$. We use vRel2 for the square of the speed (the squared length of the vector $\mathbf{v}_{rel}$).

```
Needs["VisualDSolve`"];

Initialize[wind_, ReleaseAngle_, Inclination_] := (
   g = 9.8;         degree = N[Degree];      AirDensity = 1.29;
   area = 0.038;    mass = 2.0;   y0 = 1.8;  ReleaseVelocity = 25.;
   k = AirDensity * area / mass / 2;
   ReleaseAngleRadians = ReleaseAngle * degree;
   alpha = Inclination * degree;
   {p0, q0} = ReleaseVelocity *
         {Cos[ReleaseAngleRadians], Sin[ReleaseAngleRadians]};
   beta = ArcTan[y'[t]/(x'[t] - wind)];
   theta = Abs[beta - alpha]/degree;
   {drag, lift} = {Cd[theta], Cl[theta]};
   vRel2 = y'[t]^2 + (x'[t] - wind)^2;)
```

```
DiscusEqns = {
    x"[t]  ==  -k vRel2 * (drag Cos[beta] + lift Sin[beta]),
    y"[t]  ==   k vRel2 * (lift Cos[beta] - drag Sin[beta]) - g };
```

We now use SecondOrderPlot; by specifying x and y to be the display variables, and setting AspectRatio to Automatic, we get a true picture of the flight of the discus under the given conditions. For our first simulation we consider a 10-meter-per-second-headwind, and use some carefully chosen release angles (the choice of angles is explained in section 13.4). By using StayInWindow and saving the last point, we can, after the plot is obtained, examine LastPoint to determine the exact length of the throw.

```
Initialize[-10, 31.95, 15.46];

headwind = SecondOrderPlot[DiscusEqns, {x[t], y[t]},
            {t, 0, 5}, {x, 0, 80}, {y, 0, 20},
            InitialValues -> {0, y0, p0, q0},
            StayInWindow -> True, SaveLastPoint -> True,
            FastPlotting -> True, AspectRatio -> Automatic]
```

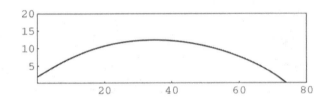

```
{-Graphics-, {{0, 0, 1.8, 21.21275537453759, 13.22947502435476,
    {0., 4.35449}}}}
```

Note how the window exit data in the output contains the initial values and the time interval corresponding to when the window was exited. Thus the flight lasted 4.35 seconds. But we want the distance, which was stored in LastPoint.

```
LastPoint
```

$$\{74.0239, 1.90742\ 10^{-9}, 11.801, -9.37268\}$$

This output shows x, y, x', and y', in that order; it is the first entry that interests us, so we conclude that the horizontal distance traveled is just over 74 meters. As the next computation shows, with a tailwind, and optimized flight angles, the attainable distance is a mere 63 meters; the difference is over 30 feet.

```
Initialize[10, 41.10, 54.04];

tailwind = SecondOrderPlot[DiscusEqns, {x[t], y[t]},
            {t, 0, 5}, {x, 0, 80}, {y, 0, 20},
            InitialValues -> {0, y0, p0, q0}, StayInWindow -> True,
            SaveLastPoint -> True, FastPlotting -> True,
            AspectRatio -> Automatic];
```

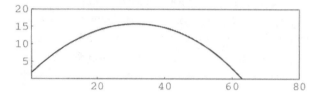

```
LastPoint[[1]]
```

63.0108

We can superimpose the two flight paths for a vivid comparison.

```
Show[headwind[[1]], tailwind[[1]]];
```

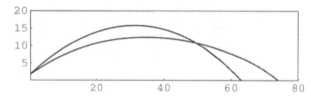

13.4 ▪ The Best Throwing Angles

It remains to discuss, for a given wind, how to find the two angles that are controlled by the thrower so as to maximize the distance achieved. An efficient way to do this is to define a distance function that returns the distance for two given angles, and then use *Mathematica*'s built-in optimization routine, FindMinimum, to find the best angles. FindMinimum uses a very efficient numerical algorithm, called the conjugate gradient method, for finding a local minimum of a function of many variables. We want a maximum, so we just minimize the negative distance.

We computed some distances in section 13.3, but that method, where we forced the solver to stay in the window (whose bottom is the ground) and then examined the last data point, is inefficient because the differential equation solver is constantly checking to see if the window has been exited. Speed is important (we will need to find hundreds of distances), so we instead compute the solution past the grounding-out point and use FindRoot to get the *x*-intercept.

```
FlightDistance[wind_, ReleaseAngle_, Inclination_] := (
   Initialize[wind, ReleaseAngle, Inclination];
   SecondOrderPlot[DiscusEqns,
      {x[t], y[t]}, {t, 0, 5}, {x, 0, 80}, {y, 0, 20},
      InitialValues -> {0, y0, p0, q0},
      DisplayFunction -> Identity, SolutionName -> "FlightData"];
   FlightData[[1]] /. FindRoot[FlightData[[2]] == 0, {t, 4}])
```

```
FlightDistance[10, 41.10, 54.04]
```
```
63.0183
```

Because the root-finding method is quite precise, we get a slightly different answer than when we used LastPoint on page 184. Now we are ready to use FindMinimum to to discover the angles in the case of a 10-meter-per-second tailwind. The numerical optimization requires solving the differential equation over 100 times (to discover this, one can insert a counter into the first argument of FindMinimum), and so takes some time (about 75 minutes on a PowerMac 8100).

```
FindMinimum[-FlightDistance[10, r, a],
   {r, {5, 60}}, {a, {5, 60}}, PrecisionGoal -> 2]
```
```
{-63.0184, {r -> 41.0997, a -> 54.0405}}
```

```
FindMinimum[-FlightDistance[-10, r, a],
   {r, {5, 60}}, {a, {5, 60}}, PrecisionGoal -> 2]
```
```
{-74.0241, {r -> 31.9505, a -> 15.4596}}
```

These results explain the angles that were used in section 13.3.

Several other questions about discus flight are worthy of investigation. For example, one can try to take the spin of the discus into account. And, of course, in the real world one would want to know how sensitive flight-length is with respect to small (or not so small) departures from the optimal angles. We refer the reader to [Fro] for further discus discussion.

Chapter 14

A Double Pendulum

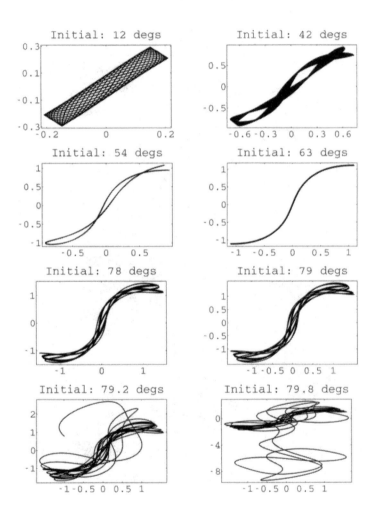

Overview

A double pendulum leads to a fairly complicated differential equation. Here we present the equations, show how to generate solutions in a variety of forms, and show how to create a movie that shows the actual pendulum swinging.

14.1 Shoulder to Elbow

By a double pendulum, we mean a massless rod (of length l_1 with mass m_1 at its end) that swings freely from a fixed frictionless pivot and has another swinging rod (length l_2 and mass m_2 at its end) dangling from the free end of the first rod (see figure). We use θ to denote the angle that the first rod makes with the vertical, while ρ is the angle made by the second rod with the vertical. And we measure y in the downward direction.

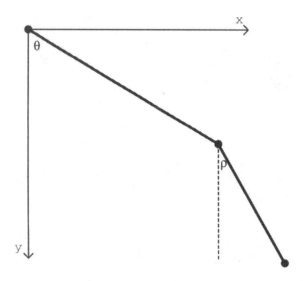

The following code defines the ends of the rods in terms of θ, ρ, and the lengths. We use a delayed assignment so that we can change mass and length later.

```
P[1] := l[1] {Sin[theta[t]], Cos[theta[t]]};
P[2] := l[2] {Sin[rho[t]], Cos[rho[t]]} + P[1];
```

Next we can define—again using delayed assignments—the kinetic energy (T) and potential energy (U) of the system. First some auxiliary definitions: we use `norm` to get the speed of the points.

```
g = 9.8;

norm[v_] := Sqrt[v . v]
speed[p_] := norm[D[p, t]]

T := 1/2 (m[1] speed[P[1]]^2 + m[2] speed[P[2]]^2) //Simplify;
U := - m[1] g P[1][[2]] - m[2] g P[2][[2]] //Simplify;
```

The Lagrangian, $T - U$, can be used to get a second-order system for θ and ρ. We do not give the details here (consult a mechanics text that discusses Lagrangian mechanics), but one can show that a certain equality of the derivatives of the Lagrangian with respect to θ, ρ, θ', and ρ' yields the appropriate differential equations. Of course, we use *Mathematica* to perform the differentiation.

```
Lagrangian = T - U;
rawEqns = {
   D[Lagrangian, theta'[t], t] - D[Lagrangian, theta[t]] == 0,
   D[Lagrangian,   rho'[t], t] - D[Lagrangian,   rho[t]] == 0}

{9.8 l[1] m[1] Sin[theta[t]] + 9.8 l[1] m[2] Sin[theta[t]] -
     l[1] l[2] m[2] Sin[rho[t] - theta[t]] rho'[t] theta'[t] +

   (2 l[1]  m[1] theta"[t] +
       m[2] (-2 l[1] l[2] Sin[rho[t] - theta[t]] rho'[t]
             (rho'[t] - theta'[t]) +
           2 Cos[rho[t] - theta[t]] l[1] l[2] rho"[t] +
              2
           2 l[1]  theta"[t])) / 2 == 0,
  9.8 l[2] m[2] Sin[rho[t]] +
     l[1] l[2] m[2] Sin[rho[t] - theta[t]] rho'[t] theta'[t] +
     (m[2] (-2 l[1] l[2] Sin[rho[t] - theta[t]]
                                                     2
           (rho'[t] - theta'[t]) theta'[t] + 2 l[2]  rho"[t] +
           2 Cos[rho[t] - theta[t]] l[1] l[2] theta"[t])) / 2 == 0}
```

In order to use `SecondOrderPlot` we have to get the equations into the form having the second derivatives isolated on the left; `rawEqns` consists of two equations, each of which involves `theta"[t]` and `rho"[t]`. So we must solve for `theta"[t]` and `rho"[t]`, which is straightforward. `Thread` can be helpful; it threads a function across its arguments.

```
Thread[f[{x, y}, {3, 4}]]
{f[x, 3], f[y, 4]}
```

We can thread across an equals sign, since `==` is short for `Equal`.

```
Thread[{x, y} == {3, 4}]
{x == 3, y == 4}
```

Now we can use `Solve` to get the standard form we want.

```
generalEqns = Thread[{theta"[t], rho"[t]}  ==
  ({theta"[t], rho"[t]} /.
  First[Solve[rawEqns, {theta"[t], rho"[t]}]])]
```

$$\{theta"[t] == -((1[2]^2 m[2]$$
$$(9.8\ 1[1]\ m[1]\ Sin[theta[t]] +$$
$$9.8\ 1[1]\ m[2]\ Sin[theta[t]] -$$
$$1[1]\ 1[2]\ m[2]\ Sin[rho[t] - theta[t]]\ rho'[t]^2) -$$
$$Cos[rho[t] - theta[t]]\ 1[1]\ 1[2]\ m[2]$$
$$(9.8\ 1[2]\ m[2]\ Sin[rho[t]] +$$
$$1[1]\ 1[2]\ m[2]\ Sin[rho[t] - theta[t]]\ theta'[t]^2)) /$$
$$(-(Cos[rho[t] - theta[t]]^2\ 1[1]^2\ 1[2]^2\ m[2]^2) +$$
$$1[2]^2\ m[2]\ (1[1]^2\ m[1] + 1[1]^2\ m[2]))),$$
$$rho"[t] == -((9.8\ 1[2]\ m[2]\ Sin[rho[t]] +$$
$$1[1]\ 1[2]\ m[2]\ Sin[rho[t] - theta[t]]\ theta'[t]^2) /$$
$$(1[2]^2\ m[2])) + (Cos[rho[t] - theta[t]]\ 1[1]$$
$$(1[2]^2\ m[2]\ (9.8\ 1[1]\ m[1]\ Sin[theta[t]] +$$
$$9.8\ 1[1]\ m[2]\ Sin[theta[t]] -$$
$$1[1]\ 1[2]\ m[2]\ Sin[rho[t] - theta[t]]\ rho'[t]^2) -$$
$$Cos[rho[t] - theta[t]]\ 1[1]\ 1[2]\ m[2]$$
$$(9.8\ 1[2]\ m[2]\ Sin[rho[t]] +$$
$$1[1]\ 1[2]\ m[2]\ Sin[rho[t] - theta[t]]\ theta'[t]^2)) /$$
$$(1[2]\ (-(Cos[rho[t] - theta[t]]^2\ 1[1]^2\ 1[2]^2\ m[2]^2) +$$
$$1[2]^2\ m[2]\ (1[1]^2\ m[1] + 1[1]^2\ m[2])))\}$$

At last we can use `SecondOrderPlot` to get a picture of how the θ and ρ values change in time.

We can look at the individual plots against time. We wish to use `FastPlotting` throughout this chapter, so we make it the default. We start with the apparently simple case where the upper weight is very large and the only initial perturbation is a 3° offset for the lower weight.

```
Needs["VisualDSolve`"];

SetOptions[PhasePlot, FastPlotting -> True];
SetOptions[SystemSolutionPlot, FastPlotting -> True];


sizes = {1[1] -> 1, 1[2] -> 1, m[1] -> 100, m[2] -> 1};
eqns = generalEqns /. sizes;
```

```
SecondOrderPlot[eqns, {theta[t], rho[t]}, {t, 0, 35},
   PlotVariables -> {{theta, rho}},
   PlotStyle -> {{AbsoluteThickness[2.2]}, {}},
   InitialValues -> {0, 3 Degree, 0, 0}, AxesLabel -> {time, degrees},
   MaxSteps -> 2000, PlotRange -> All, AxesOrigin -> {0, -3.2 Degree},
   Ticks -> {Automatic, Table[{a Degree, a}, {a, -3, 3, 1}]},
   PlotLabel ->
      FontForm["theta = thick; rho = thin", {"Courier", 9}]];
```

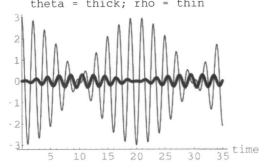

Readers who have never seen a double pendulum are encouraged to examine these graphs and try to imagine what the swinging action will be like. Even better: get two weights and some string and make a rudimentary double pendulum and play with it! Note how ρ, the thin curve, gives away some of its energy to θ at the beginning, only to get it back shortly thereafter. Similarly, θ swings in spurts. This behavior, called *beating*, is easily seen in the phase plane view of θ vs. ρ.

```
PendulumSpace = SecondOrderPlot[eqns, {theta[t], rho[t]}, {t, 0, 23},
   {theta, -1/3 Degree, 1/3 Degree}, {rho, -3.5 Degree, 3.5 Degree},
   InitialValues -> {0, 3 Degree, 0, 0},
   AxesLabel -> {"theta (degrees)", rho},
   Ticks -> {Table[{y Degree, Chop[y]}, {y, -0.2, 0.2, 0.2}],
             Table[{y Degree, y}, {y, -3, 3}]},
   SolutionName -> "BeatingPendulum", MaxSteps -> 2000];
```

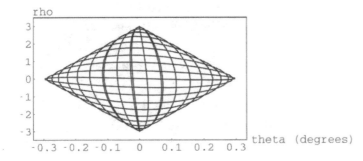

A natural question is whether there is periodicity here. We can compute the solution out farther. The result seems to show that there is not exact periodicity for this initial condition.

```
SecondOrderPlot[eqns, {theta[t], rho[t]}, {t, 0, 110},
    {theta, -1/3 Degree, 1/3 Degree}, {rho, -3.5 Degree, 3.5 Degree},
    InitialValues -> {0, 3 Degree, 0, 0}, AxesLabel -> {theta, rho},
    MaxSteps -> 2000,
    Ticks -> {Table[{y Degree, Chop[y]}, {y, -0.3, 0.3, 0.1}],
    Table[{y Degree, y}, {y, -3, 3}]}, SaveLastPoint -> True];
```

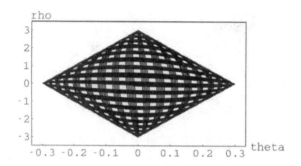

Of course, we should pay attention to the question of the accuracy of the solution. Two approaches come to mind. The small perturbation method, discussed in detail in Chapter 15, asks for the effect of a small perturbation on the initial conditions. The perturbation should be about 10^{-6}, as explained in Chapter 15. When we do that here we get only a small change in the last point, and that is good evidence that the solution is correct.

But here we have access to another method of checking. The double pendulum equations are Hamiltonian, with Hamiltonian function (total energy) given by $T + U$. So we can examine the value of $T + U$ on the orbit just generated; since it takes a long time to differentiate the complicated interpolating functions, we content ourselves with checking the energy at the starting and ending points (the final values were saved in `LastPoint`, thanks to the `SaveLastPoint` option). First we compute the energy at the start, then the energy at the end, using the threading technique to get the four replacement rules from the list of four values.

```
T + U /. {Thread[Rule[{theta[t], rho[t], theta'[t], rho'[t]},
                {0, 3 N[Degree], 0, 0}]],
        Thread[Rule[{theta[t], rho[t], theta'[t], rho'[t]},
                LastPoint]]} /. sizes
```

```
{-999.587, -999.587}
```

The results are good and give us faith in the solution out to 110 seconds. When we look at greater amplitudes, however, the energy conservation will not last this long.

14.2 ■ Linearization

One can often learn something by using the linearization of the system, especially when the angles are known to stay small. We illustrate how to do that. First we turn the two second-order differential equations into a system (masses and lengths are still set as in section 14.1).

```
pendSystem = ToSystem[eqns, {theta[t], rho[t]}, t,
  AuxiliaryVariables -> {u, v}]
```

```
{theta'[t] == u[t], rho'[t] == v[t],
   u'[t] == -((9.8 Cos[rho[t] - theta[t]] Sin[rho[t]] -
       989.8 Sin[theta[t]] +

       Cos[rho[t] - theta[t]] Sin[rho[t] - theta[t]] u[t]² +

       Sin[rho[t] - theta[t]] v[t]²) /

       (-101. + Cos[rho[t] - theta[t]]²)),
   v'[t] == -((-989.8 Sin[rho[t]] +
       989.8 Cos[rho[t] - theta[t]] Sin[theta[t]] -

       101. Sin[rho[t] - theta[t]] u[t]² -

       Cos[rho[t] - theta[t]] Sin[rho[t] - theta[t]] v[t]²) /

       (-101. + Cos[rho[t] - theta[t]]²))}
```

Next we get the Jacobian matrix by doing 16 differentiations. We use `Outer`, which takes all combinations as we illustrate with a simple example.

```
Outer[func, {a, b}, {d, e, f}]
```

```
{{func[a, d], func[a, e], func[a, f]},
  {func[b, d], func[b, e], func[b, f]}}
```

```
unks = {theta[t], rho[t], u[t], v[t]};
```

```
MatrixForm[A = Outer[D, Map[Last, pendSystem], unks] /.
  Thread[Rule[{theta[t], rho[t], u[t], v[t]},
  {0, N[3 Degree], 0, 0}]] ]
```

0	0	1	0
0	0	0	1
-9.89746	0.0974551	0	0
9.88362	-9.88362	0	0

Now the linearized system has just A . unks on the right side. We don't show the image of the solution since it is identical to that on page 191.

```
SystemSolutionPlot[Thread[D[unks, t]  == A . unks], unks,
  {t, 0, 35},
  PlotVariables -> {{theta, rho}},
  PlotStyle -> {{AbsoluteThickness[2.2]}, {}},
  InitialValues -> {0, 3 Degree, 0, 0},
  AxesLabel -> {time, degrees}, MaxSteps -> 2000,
  PlotRange -> All,
  Ticks -> {Automatic, Table[{a Degree, a}, {a, -3, 3, 1}]},
  PlotLabel -> FontForm["theta = thick; rho = thin", {"Courier", 10}]]
```

One advantage is that we can obtain a symbolic solution. We use ComplexExpand to turn the complex exponentials into trig functions.

```
symbolicSol = ({theta[t], rho[t]} /. First[
  DSolve[Join[Thread[D[unks, t]  == A . unks],
    {theta[0] == 0, rho[0] == 3 Degree, theta'[0] == 0, rho'[0] == 0}],
      unks, t]]  //ComplexExpand //Chop)
```

```
{0.0523599 (0.0496482 Cos[2.98481 t] - 0.0496482 Cos[3.29727 t]),
  0.0523599 (0.503523 Cos[2.98481 t] + 0.496477 Cos[3.29727 t])}
```

Note that these functions, assuming the real numbers are nothing special, are not periodic, though they might be called almost periodic. The graph on page 191 indicated that there was a potential period of length 19. Here is how we can find a more exact value: we try to find a *t*-value for which both summands in the closed form are close to being periodic. The function err[z] measures the distance from z to the nearest multiple of 2π.

```
a = 2.98481; b = 3.29727;
err[z_] := Abs[N[z - 2 Pi Round[z / (2 Pi)]]]
FindMinimum[err[a t] + err[b t], {t, {18.5, 21}}]
```

```
{0.329035, {t -> 19.0557}}
```

While the discrepancy from periodicity (0.329) seems large, keep in mind that there are small numbers that multiply the cosine functions. The following evaluation shows that both the numerical solution and the closed form are near the initial conditions $\{0, 0.0523599\}$ at $t = 19.0557$. One should, of course check θ' and ρ' as well.

```
symbolicSol /. t -> 19.0557
BeatingPendulum /. t -> 19.0557
3 Degree//N
```

```
{-0.000139383, 0.0509463}
{-0.000147158, 0.0508564}
0.0523599
```

To summarize, we have looked at the linearization of the double pendulum and observed that it is close to being periodic at $t = 19.0557$. Because the initial angles are small, the linearization is a reasonable approximation to the nonlinear system.

14.3 ■ Bringing the Pendulum to Life

Now it is a simple matter to make a movie that shows the swinging pendulum. We first define rod1 and rod2 to be expressions in t that give the positions of the rod-ends; because y was taken in the downward direction, we need to flip the y-coordinate (multiplication by {1, -1}) to conform to the usual coordinate system. We use the fact that BeatingPendulum contains the two interpolating functions we need (stored by the code on page 191). We also tweak the aspect ratio, for otherwise the small amplitude of the upper rod does not show up on the screen; thus the images are not true, but exaggerate the horizontal variation. (We have displayed only one frame from the animation.)

```
{rod1, rod2} = {P[1] * {1, -1}, P[2] * {1, -1}} /.
   sizes /. {theta[t] -> BeatingPendulum[[1]],
             rho[t] -> BeatingPendulum[[2]]};


Do[Show[Graphics[{AbsoluteThickness[2],
   Rectangle[{-0.05, 0}, {0.05, 0.07}],
     Line[{{0, 0}, rod1, rod2}],
     AbsolutePointSize[20], Point[rod1],
     AbsolutePointSize[5], Point[rod2]}],
   PlotRange -> {{-0.3, 0.3}, {-2.3, .2}},
   PlotLabel -> StringForm["`` seconds", N[t, 4]],
   AspectRatio -> 1/3,
   DefaultFont -> {"Courier-Bold", 12}],  {t, 0, 21, 1/4}]
```

0 seconds

In fact, it is not hard to make a movie that combines the swinging pendulum with the motion of the point on the trajectory in the $\theta - \rho$ plane. (Again, just one frame from the animation is displayed here.)

```
pendImage[time_] := Graphics[{
    AbsoluteThickness[1.5], AbsolutePointSize[6],
    Line[pts = {{0, 0}, rod1 /. t-> time, rod2 /. t -> time}],
    Map[Point, pts]},
  PlotRange -> {{-0.3, 0.3}, {-2.3, .2}},
  AspectRatio -> Automatic];

orbitAndPoint[time_] := Graphics[{PendulumSpace[[1]],
  {AbsolutePointSize[5], Point[BeatingPendulum /. t -> time]}}]

Do[Show[GraphicsArray[{orbitAndPoint[t], pendImage[t]}]],
  {t, 0, 3, 0.25}]
```

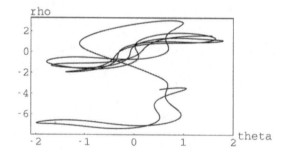

14.4 ■ Chaos, and What Happens on the Way

If we look at a more energetic system, then the patterns we saw for small initial values totally disappear. Here is a view of θ and ρ if we start with both angles horizontal. Note how ρ heads south. This indicates multiple clockwise spins of the lower arm! We only run t out to 14 since the perturbation method (discussed in Chapter 15) indicates that going any farther leads to unreliable results.

```
sizes = {l[1] -> 1, l[2] -> 1, m[1] -> 1, m[2] -> 1};
eqns = generalEqns /. sizes;

SecondOrderPlot[eqns, {theta[t], rho[t]},
  {t, 0, 14}, {theta, -2.1, 2}, {rho, -8, 3.3},
  InitialValues -> {Pi/2, Pi/2, 0, 0},
  MaxSteps -> 3000, AxesLabel -> {theta, rho}];
```

It is instructive to make a movie that shows the progression from the orderly behavior at small amplitudes to the apparently chaotic behavior just seen. It is important to note that the use of the word "chaos" in this context has a precise meaning; from a strict mathematical viewpoint, it is most desirable to prove that chaos exists, but this has been done only in a few cases. It is not yet proved that the double pendulum is truly chaotic. As for chaos's meaning, a rough definition is that a trajectory is chaotic if it exhibits sensitive dependence on initial conditions (nearby starting points lead to very different orbits), and if it, and orbits with nearby initial conditions, have no fixed points or periodic behavior. See [Str, chap. 9] for more precision and for the closely related concept of a strange attractor.

While the pendulum system has four dependent variables, remember that it is Hamiltonian, and so, since the energy function must stay constant, the fourth variable can be determined from the other three. Thus we can get a complete picture by looking at the 3-dimensional trajectory of the first three variables. This is not hard to do, but we stick to two-dimensional views here since they are equally informative and quicker to generate.

`SecondOrderPlot` requires the user to have an idea of the viewing window. Here, though, we wish to see the whole trajectory, and the space it spans will change as the amplitude increases. Thus we resort to the following strategy: We define `pend` to generate the solution, which we get by calling `NDSolve` (we could also accomplish this by using the `SaveSolution` option to `SecondOrderPlot`). The we get the points in the two interpolating functions, transpose them, and connect the dots. This is considerably faster than using `Plot`. Using `Scan` allows us to generate a movie with more frames in the region of interest: the transition to chaos near 80°.

```
pend[amp_] := {theta[t], rho[t]} /. First[
   NDSolve[Join[eqns, {theta[0] == amp Degree,
      rho[0] == amp Degree, theta'[0] == 0, rho'[0] == 0}],
      {theta[t], rho[t]}, {t, 0, 30}, MaxSteps -> 3000]]


Scan[(
   amp = #;
   Show[Graphics[{AbsoluteThickness[0.5],
         Line[Transpose[Map[GetPts, pend[amp]]]]}],
      PlotLabel ->
         StringForm["Initial amplitudes: `` degrees", amp],
      PlotRange -> All, Frame -> True])&,
   Join[Range[12, 78, 3], Range[78.2, 81, 0.2]]]
```

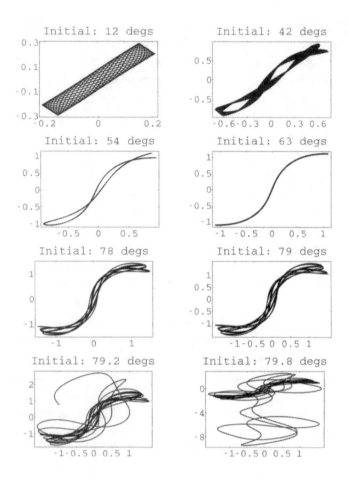

There are several interesting aspects to this sequence of trajectories. Most noteworthy is that there seems to be a periodic solution near 63°. Moreover, the trajectory looks perfectly symmetric. More careful work indicates that this occurs at 63.3°. We should look at the third variable, θ', as well. Because `SecondOrderPlot` can handle only two viewing iterators, we convert the equation to a system and use `PhasePlot` to generate a three-dimensional orbit. The $\theta - \rho$ symmetry carries over to u, which represents θ' and takes on both positive and negative values.

```
pendFirstOrder = ToSystem[
    eqns, {theta[t], rho[t]}, t, AuxiliaryVariables -> {u, v}];

PhasePlot[pendFirstOrder, {theta[t], rho[t], u[t], v[t]}, {t, 0, 30},
    {theta, -1.5, 1.5}, {rho, -2, 2}, {u, -3, 3},
    InitialValues -> {63.3 Degree, 63.3 Degree, 0, 0},
    PlotLabel -> "63.3 degrees", MaxSteps -> 2000];
```

63.3 degrees

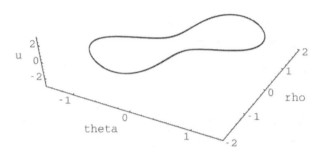

It is useful to look at the individual plots against time as well. The following pair of graphs shows two periods, with θ and θ' on the left and ρ and ρ' on the right.

```
SystemSolutionPlot[pendFirstOrder,
  {theta[t], rho[t], u[t], v[t]}, {t, 0, 2.80134 * 2},
  InitialValues -> {63.3 Degree, 63.3 Degree, 0, 0},
  PlotVariables -> {{theta, u}, {rho, v}}];
```

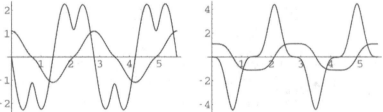

At low amplitudes the three-dimensional trajectory lives on a torus. This is because the linearization is a good approximation at low energies and the theory of KAM (Kolmogorov–Arnold–Moser) tori applies (see [Wol, Rei]). In fact, Jay Wolkowiski [Wol] shows how, with a little symbolic work, one can use *Mathematica* to derive the equations of the torus and generate it.

```
PhasePlot[pendFirstOrder, {theta[t], rho[t], u[t], v[t]},
  {t, 0, 30}, {theta, -0.3, 0.3},
  {rho, -0.3, 0.3}, {u, -0.5, 0.5},
  InitialValues -> {10 Degree, 10 Degree, 0, 0},
  PlotLabel -> "10 degrees",
  MaxSteps -> 2000];
```

10 degrees

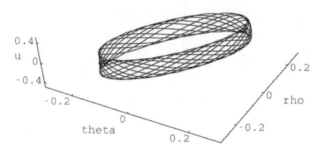

14.5 ■ Synchronized Swinging

We conclude our tour of the double pendulum with a movie that shows the effect of larger amplitudes on an actual pendulum. First we generate four solutions, with amplitudes 78°, 81°, 83°, and 85°, and store them in `swingers`. This requires the prior definition of `pendFirstOrder` as a first-order system, as was done on page 198.

```
PhasePlot[pendFirstOrder, {theta[t], rho[t], u[t], v[t]},
  {t, 0, 20}, {theta, -3, 3}, {rho, -3, 6},
  InitialValues ->
    Table[{amp Degree, amp Degree, 0, 0}, {amp, 79, 88, 3}],
  MaxSteps -> 5000,
  DisplayFunction -> Identity,
  SolutionName -> "swingers"];
```

Next we define `rod1[i]` and `rod2[i]` to contain the functions defining the pendulum-ends for the *i*th solution.

```
Do[{rod1[i], rod2[i]} = {P[1] * {1, -1}, P[2] * {1, -1}} /. sizes /.
  {theta[t] -> swingers[[i, 1]], rho[t] -> swingers[[i, 2]]},
  {i, 4}]
```

And now we can generate an 81-frame movie that shows 20 seconds worth of swinging. Note the chaotic swinging of the lower arm of the rightmost pendulum. The two middle ones will, as time goes on, show similar behavior. The printed image shows only a small section of the movie.

```
Do[Show[GraphicsArray[Table[
  Graphics[{AbsoluteThickness[2],
      {GrayLevel[0.5], Line[{{-1.5, 0}, {1.5, 0}}]},
      Line[{{0, 0}, rod1[i], rod2[i]}],
      AbsolutePointSize[6], Map[Point, {rod1[i], rod2[i]}]},
    PlotLabel -> StringForm["`` degs", 76 + 3 i],
    PlotRange -> {{-2, 2}, {-2, 1}}], {i, 4}]],
  PlotLabel -> StringForm["`` sec", N[t, 3]],
  GraphicsSpacing -> 0.08,
  PlotRange -> {{0.27, 3.8}, Automatic},
  DefaultFont -> {"Courier", 9}], {t, 0, 20, 1/4}]
```

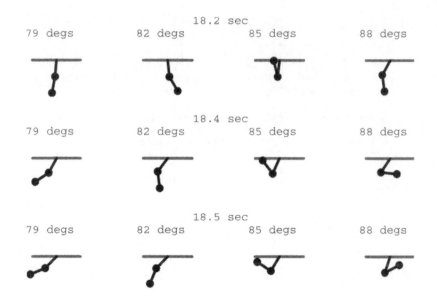

Here is a plot of ρ, with lines superimposed so we can see how many times the lower pendulum swings over the top. This graph shows one clockwise flip and three counterclockwise flips.

```
Plot[Evaluate[{swingers[[4, 2]], -Pi, Pi, 3 Pi, 5 Pi}],
   {t, 0, 20},
   PlotStyle -> Prepend[Table[AbsoluteDashing[{2, 4}], {4}], {}]];
```

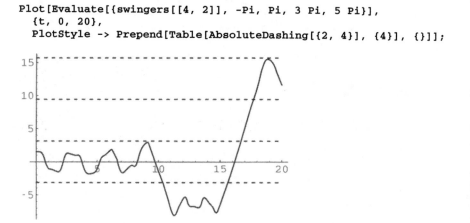

Chapter 15

The Duffing Equation

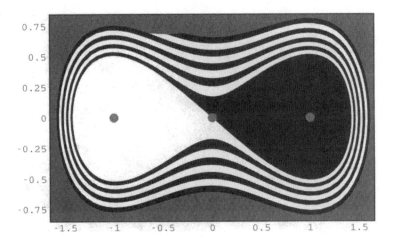

Overview

The Duffing equation is a much-studied differential equation arising from a forced system (typically an oscillator subject to a periodic force; for physical details see [BCB] or [Str] and for more on the fascinating mathematics of this equation see [GH, chap. 2]). Duffing orbits exhibit a fascinating variety of behaviors typical of chaos. In this chapter we show some interesting orbits, and discuss ways to recognize and deal with sensitive situations. The techniques that can help us recognize an inaccurate solution are quite general, and should be used whenever a software user seeks evidence regarding the accuracy of the output of a numerical algorithm.

15.1 ■ A Stable Example

The Duffing equation is $x'' - ax + bx^3 + cx' = d\cos(et)$. We begin by defining the equation and looking at some simple cases. Because e, the constant that sets the period of the forcing term, is often taken to be 1, we set that as a default.

```
Duffing[a_, b_, c_, d_, e_:1] :=
    x"[t] - a x[t] + b x[t]^3 + c x'[t] == d Cos[e t]
```

We begin with an unforced case ($d = 0$).

```
Needs["VisualDSolve`"];

SecondOrderPlot[Duffing[0.5, 0.5, 0.015, 0],
    x[t], {t, 0, 50}, {x, -1.7, 1.7}, {x', -0.65, 0.65},
    InitialValues -> {{1, 0.54}, {1, 0.56}},
    AxesLabel -> {x, x'}, NullclineShading -> True,
    ShowEquilibria -> True, PlotPoints -> 50];
```

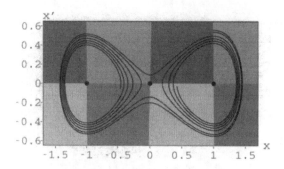

Note how the nearby initial values lead to orbits that spiral toward different equilibria. We can use `ProjectionPlot3D` to get a three-dimensional view of one of these orbits, together with its x, y, and x-y projections. The output of the next command is in color plate 21.

```
SecondOrderPlot[Duffing[0.5, 0.5, 0.015, 0],
    x[t], {t, 0, 60}, {x, -2, 2}, {x', -1, 1},
    InitialValues -> {1, 0.54}, PlotPoints -> 200,
    MaxSteps -> Infinity, ParametricPlotFunction -> ProjectionPlot3D,
    AxesLabel -> {x, x', t}];
```

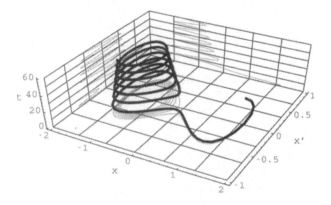

There is no indication of instability in this orbit, but we may as well discuss now a method that can be used to try to detect bad solutions. Sophisticated numerical algorithms try to estimate the error at each step; then they do additional work until this error is below a prescribed tolerance. A typical goal, and this is the default for *Mathematica*'s NDSolve, is to attempt to get this local error less than 10^{-6} (in a relative sense, of course). This means that, after the very first step, one can expect an error of 10^{-6} to be introduced. We can anticipate this error by perturbing the initial conditions by 10^{-6}. In short, a good general rule is to perturb the initial conditions by an amount on the order of the local truncation error. Let's do it.

```
e = 10^-6;
SecondOrderPlot[Duffing[0.5, 0.5, 0.015, 0],
    x[t], {t, 0, 100}, {x, -1.7, 1.7}, {x', -0.65, 0.65},
    InitialValues -> {{1, 0.54}, {1 + e, 0.54 + e}},
    MaxSteps -> 2000, PlotPoints -> 50, SaveLastPoint -> True];
```

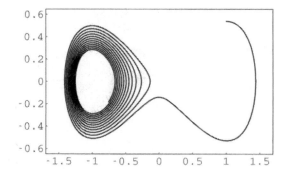

```
LastPoint
```

{{-0.757735, -0.182259}, {-0.75769, -0.182227}}

The results differ by an amount similar to the perturbation, which indicates the absence of a large error-propagation effect. A large discrepancy, on the other hand, would indicate that the error propagation was large, and so the results cannot be trusted.

It is natural to wonder which initial conditions lead to which equilibrium points. Such computations of "basins of attraction" can be very time-consuming when the basins are complicated, but they do lead to very striking images, since the basins of attraction often have fractal boundaries (see section 15.4). In the present case, the system is autonomous and one can begin to investigate the basins by simply drawing lots of orbits and checking where they go.

The code that follows uses the auxiliary function GetPts from the package, which turns an interpolating function into the list of interpolating points. We can look at the last point to see which of the two equilibria has attracted the orbit, and therefore which color to use. The curve function that we now define does all this; that is, it takes an `InterpolatingFunction` and returns an appropriately colored `Line` object.

```
curve[pair_] := (temp = Transpose[Map[GetPts, pair]];
   {{If[temp[[-1, 1]] > 0, Black, White], Line[temp]}})
```

We now set some parameters so that we get 21 orbits. These parameters were chosen after some trial and error.

```
ymin = 0.555; ymax = 0.6175;
s = (ymax - ymin)/20;
{a, b} = {1.6, 0.73};

SecondOrderPlot[Duffing[0.5, 0.5, 0.015, 0],
   x[t], {t, 0, 100}, {x, -a, a}, {x', -b, b},
   InitialValues -> Table[{1, y}, {y, ymin, ymax, s}],
   DisplayFunction -> Identity,
   SolutionName -> "DuffingOrbits", MaxSteps -> 800];
```

Finally, we apply curve to the functions stored in `DuffingOrbits` to get an informative image.

```
Show[Graphics[{AbsoluteThickness[0.4],
   Map[curve, DuffingOrbits]},
   DisplayFunction -> $DisplayFunction,
   Prolog -> {{GrayLevel[0.4], Rectangle[{-a, -b}, {a, b}]},
             {Thickness[0.004],
              Line[{{-a, -b}, {a, -b}, {a, b}, {-a,b}, {-a,-b}}]}},
   Frame -> True, PlotRange -> {{-a, a}, {-b, b}}];
```

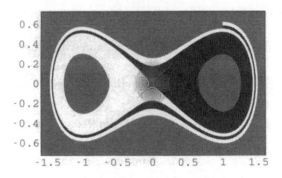

This image shows that the two basins of attraction are intertwined, with smooth boundaries. But let's not forget about the separatrices: orbits that go exactly to the saddle equilibrium at the origin. In fact, we can generate the two separatrices by considering two initial values, one just above and one just below the origin. Then a little graphics programming using the points on the separatrices will yield two polygon objects that capture the basins perfectly.

First we generate the separatrices. Note how the initial conditions specify that the initial value is at $t = 0$, but the plotting intervals are taken to be negative. Thus we are reversing time to get these orbits.

```
SecondOrderPlot[Duffing[0.5, 0.5, 0.015, 0],
    x[t], {t, -66.7, 0}, {x, -1.7, 1.7},  {x', -0.83, 0.83},
    InitialValues -> {{0, 0, -0.001}, {0, 0, 0.001, {-61.9, 0}}},
    PlotStyle -> {{Black}, {White}},
    FastPlotting -> True,
    WindowShade -> GrayLevel[0.45],
    SolutionName -> "DuffingSeparatrices",
    MaxSteps -> 2000];
```

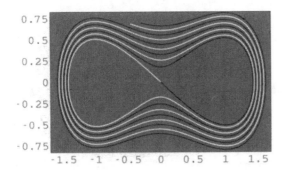

The programming that follows is a little intricate, since we must get the exact points to match up in order to get both `Polygon` objects. In particular, we need to locate which point in the black interpolating function (in the preceding image) is near (0.64, 0.69). It turns out to be the 86th point, but since we must reverse this data set (473 points) to link the polygon properly, we take the first 388 entries of `Reverse[highData]`). The color image is in color plate 22.

```
{a, b} = {1.66, 0.85};
highData = Transpose[Map[GetPts, DuffingSeparatrices[[1]]]];
lowData  = Transpose[Map[GetPts, DuffingSeparatrices[[2]]]];

Show[Graphics[{
    {Blue, Polygon[Join[Reverse[highData], lowData]]},
    {Yellow, Polygon[
      Join[Reverse[lowData], Reverse[Take[Reverse[highData], 388]]]]},
    {AbsolutePointSize[7], GrayLevel[0.5], Red,
      Map[Point, {{-1, 0}, {1,0}, {0, 0}}]}}],
    Frame -> True, FrameTicks -> {Automatic, Automatic, None, None},
    Prolog -> {{GrayLevel[0.4], Rectangle[{-a, -b}, {a, b}]},
      {Thickness[0.004],
        Line[{{-a, -b}, {a, -b}, {a, b}, {-a, b}, {-a, -b}}]}},
    PlotRange -> {{-a, a}, {-b, b}}];
```

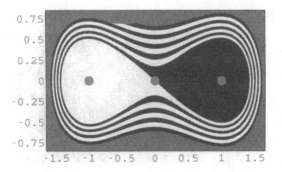

This image shows the basins quite nicely. Of course, generating such an image for a general autonomous system might well involve some graphics tricks peculiar to the situation. But the ideas above should work in many other autonomous cases as well. Nonautonomous cases are a different story!

15.2 ▪ The General Duffing Equation

We now study the impact of a forcing term. And we continually use the perturbation idea just discussed as a way of monitoring the likelihood that the results are accurate. The forced system—the forcing term is $0.3 \cos t$—is not autonomous, so we cannot look at nullclines or equilibrium points.

```
SecondOrderPlot[Duffing[1, 1, 0.15, 0.3],
    x[t], {t, 0, 6}, {x, -2.2, 2.2}, {x', -2.2, 2.2},
    InitialValues -> {{0.6, 1.3}, {0.6 + e, 1.3 + e}},
    MaxSteps -> 2000, FastPlotting -> True];
```

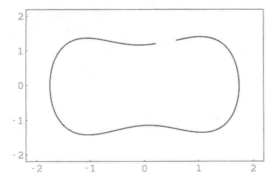

```
e = 10^-6;
SecondOrderPlot[Duffing[1, 1, 0.15, 0.3],
   x[t], {t, 0, 40}, {x, -2.2, 2.2}, {x', -2.2, 2.2},
   InitialValues -> {{0.6, 1.3}, {0.6 + e, 1.3 + e}},
   MaxSteps -> 2000, FastPlotting -> True,
   PlotPoints -> 50, SaveLastPoint -> True];
```

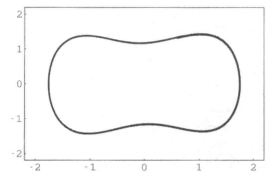

In this case the solution seems very close to being periodic. To examine the result of the perturbation, we first define a distance function.

```
distance[{p_, q_}] := Sqrt[(p - q) . (p - q)]
LastPoint
```

```
{{0.379919, -1.19993}, {0.379918, -1.19991}}
```

```
distance[LastPoint]
```

```
0.0000167267
```

Again, the results are good in that the distance between the last points is only a little larger than the perturbation. Of course, this is not a proof, but it is good evidence. What would happen if we perturbed by 10^{-3}, or 10^{-12}? We will leave such computations as exercises. But because the precision goal is only 6 digits, a 6-digit error will arise at the first step anyway, so a 10^{-12} perturbation will yield a discrepancy similar to that for a 10^{-6} perturbation. In section 15.5 we will discuss how and when to change the precision goal.

Now we take the same Duffing equation but modify the initial value.

```
SecondOrderPlot[Duffing[1, 1, 0.15, 0.3],
   x[t], {t, 0, 100}, {x, -1.7, 1.7}, {x', -1.7, 1.7},
   InitialValues -> {-1, 1}, ShowInitialValues -> True,
   FastPlotting -> True, MaxSteps -> 2000];
```

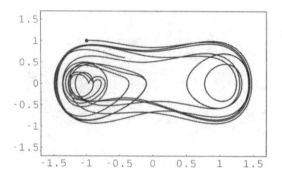

Here we see that the orbit has become quite disorganized. It looks like there are two regions that the orbit has an inclination to visit, and that it jumps from one to the other somewhat randomly. It will help to have a physical model to think about. Imagine a bead sliding on a wire in the shape illustrated. This is called a two-well model because, if a damping term is added, the bead will settle into one of the wells, the bump being an unstable equilibrium.

But if the whole apparatus is rigidly shaken from left to right with a periodic forcing function, then the bead can slide chaotically from well to well. The equations for this forced apparatus are quite similar (though not identical) to the Duffing equation (see [Str, section 12.5]). Thus the orbit just generated can be thought of as showing the two wells, with the bead moving around in one of the wells for a while and then jumping to the other one.

Of course, when the motion is this complicated, we must wonder whether the solution is really correct. As usual, we try a perturbation of 10^{-6}.

```
e = 10^-6;
SecondOrderPlot[Duffing[1, 1, 0.15, 0.3],
    x[t], {t, 0, 100}, {x, -1.7, 1.7}, {x', -1.5, 1.5},
    InitialValues -> {{-1, 1}, {-1 + e, 1 + e}},
    ShowInitialValues -> True,
    MaxSteps -> 2000,
    SolutionName -> "BadDuffing",
    PlotStyle -> {{AbsoluteThickness[3], Gray}, {}},
    SaveLastPoint -> True];
```

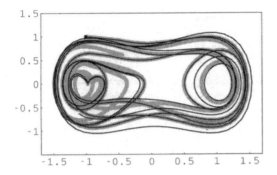

The picture is a little hard to read, but it is clear that the orbits have diverged.

```
distance[LastPoint]
```

1.38118

There are two approaches to getting more legible images. One is to plot only x against t. We could do this using SecondOrderPlot in a way that calls SystemSolutionPlot, but then the PlotStyle option only works to set styles for different plotting variables, rather than different initial values. So instead we use the fact that we have just stored the solution in BadDuffing.

```
Plot[{BadDuffing[[1, 1]], BadDuffing[[2, 1]]}, {t, 0, 100},
    PlotPoints -> 100,
    PlotStyle -> {{AbsoluteThickness[3], Gray}, {}}];
```

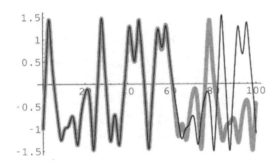

An advantage of this view is that it shows clearly the time limit for dependability. The image indicates that t-values up to about 60 are reliable; beyond that, the solutions rapidly diverge. Thus one can conclude that neither solution is accurate beyond $t = 60$.

One can also look at the phase plane image, but restrict the plotting domain. Now that we know that $t = 60$ is critical, we can look at the orbits from, say, 40 to 100. It is even more easy to read if we do it in steps of 10 seconds. The efficient way to do this is to first generate the solution out to 100 (which we have done already), and then use it to generate a few plots.

```
Do[ParametricPlot[Evaluate[BadDuffing], {t, t1, t1 + 10},
   PlotRange -> {{-1.7, 1.7}, {-1.5, 1.5}},
   Epilog -> {AbsolutePointSize[5], Map[Point, BadDuffing /. t -> t1]},
   PlotLabel -> StringForm["t from " to "", t1, t1 + 10],
   AxesOrigin -> {-1.7, -1.5},
   PlotStyle -> {{AbsoluteThickness[3], GrayLevel[0.7]}, {}}],
   {t1, 40, 90, 10}];
```

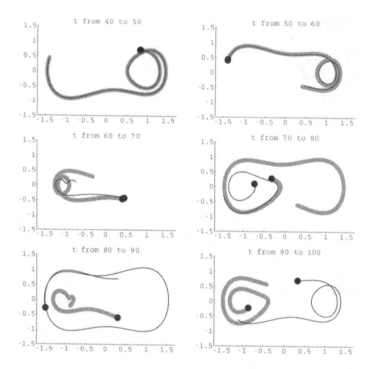

Include some overlap (by using, say, {t1, 40, 90, 5}) and you can get a movie showing the motion. Note that this output shows much more than sensitivity to initial conditions. We already saw such sensitivity, to some extent, in section 15.1. Rather, this output is telling us that the numerical method is totally unreliable, and that both orbits we are seeing are, almost certainly, incorrect.

Several books have published orbits of the Duffing equation with *t* running up to 10,000 or more. But the above computations show that, if machine precision is used (as is the case with almost all differential equations software), then the results are not correct.

15.3 ■ Flying Duffing Circles

Returning to the more modest range of *t* between 0 and 10, we can gain some understanding of why the extreme sensitivity exists by making some movies. The nullcline shading idea does not apply to a nonautonomous system, but we can apply it at a given time instant. In other words, the underlying flow field changes as a function of t (because of the $\cos t$ term), but by fixing t we can look at the instantaneous null-clines. Similarly, there are no equilibrium points, but we can look at the "instantaneous equilibrium points" (which correspond to the two wells and the bump).

So here's the plan: consider a small circle (radius = 1/10) of initial values around $(-1, 1)$, and make a movie that shows where these initial values go as t increases. In each frame we can include the instantaneous equilibria (the wells and bump) and nullcline shading. The code for generating these movies is discussed at the end of this section.

An Insensitive Case

In the insensitive case—initial values (0.6, 1.3)—we see that the thumping term does not greatly distract the flying circle. Using the two-well analogy, the ball seems to have enough energy to fly around the surface in a way that avoids the wells and the bump between, and it is unaffected by the forcing term.

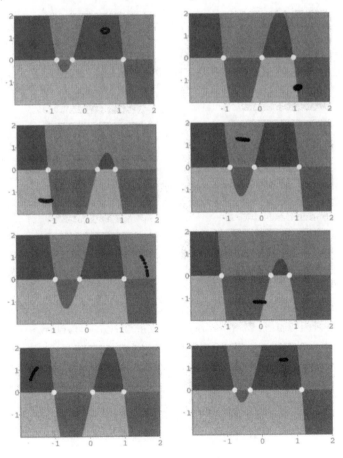

A Sensitive Case

However, if we start at $(-1, 1)$, then the nearby initial conditions do interact with the bump. The animation shows how the bump breaks up the circle, causing some orbits to visit one well while the rest visit another. And keep in mind that this behavior is occurring over the modest time range of $[0, 4\pi]$.

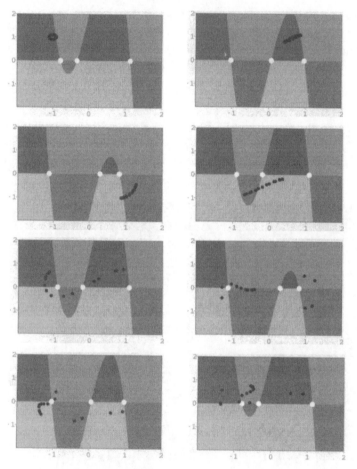

The shape of the image of the initial circle in the sensitive case is intriguing in the way that it folds back upon itself. In fact, this fractal-like folded-over shape occurs again and again in orbits of the Duffing equation, and is related to something called the Smale horseshoe map (see [Str, chap. 12]). It is strongly suspected (though not proved) that a fractal-like strange attractor exists for the Duffing equation. Briefly, this refers to a very complicated set, with a fractal nature, that attracts even chaotic orbits.

Code for Flying Duffing Circles

First generate the `InterpolatingFunctions` for sixteen different solutions and store them.

```
SecondOrderPlot[Duffing[1, 1, 0.15, 0.3], x[t],
  {t, 0, 4 Pi}, {x, -2, 2}, {x', -2, 2},
  InitialValues ->
    Table[{{-1, 1} + 0.1 {Cos[s], Sin[s]}, {s, 0, 15 Pi/8, Pi/8}],
  SolutionName -> "SixteenDuffings", MaxSteps -> 3000,
  DisplayFunction -> Identity];
```

Then define a data function that captures the sixteen positions at a given time.

```
data[tval_] := SixteenDuffings /. t -> N[tval]
```

Now generate a pseudo-nullcline plot by fixing the *t*-value at `t1`; then super-impose the flying circle using `Epilog`. Letting `t1` vary in a Do-loop gets us the movie we want.

```
Do[SecondOrderPlot[
    x"[t] == -0.15 x'[t] + x[t] - x[t]^3 + 0.3 Cos[t1],
    x[t], {x, -2, 2}, {x', -2, 2},
    NullclineShading -> True, ShowEquilibria -> True,
    EquilibriumPointStyle -> {AbsolutePointSize[5], White},
    NullclinePlotPoints -> 30,
    Epilog -> {AbsolutePointSize[3], Map[Point, data[t1]]}],
  {t1, 0, 4 Pi, 4 Pi/20}]
```

Another way of gaining appreciation of the evolution of Duffing orbits is to look at a single image that shows what happens to a circle of initial conditions. In the code that follows we look at orbits out to $t = 7$ and connect the resulting points (32 of them, from a circle of 32 initial points) with lines so as to see what happens to the initial circle. Note the folding that takes place. Note also that the area within the evolving circle seems to shrink. These behaviors are connected to the Lyapunov exponents, which can be estimated [KW].

```
SecondOrderPlot[Duffing[1, 1, 0.15, 0.3], x[t],
  {t, 0, 7}, {x, -2, 2}, {x', -2, 2},
  InitialValues -> Prepend[Table[{-1, 1} + 0.1 {Cos[s], Sin[s]},
      {s, 0, 2 Pi - Pi/32, Pi/32}], {-1, 1}],
  SolutionName -> "DuffingOrbits",
  MaxSteps -> 3000, DisplayFunction -> Identity];
```

```
main = ParametricPlot[Evaluate[First[DuffingOrbits]], {t, 0, 7},
  PlotStyle -> {{AbsoluteThickness[3], GrayLevel[0.5]}},
  DisplayFunction -> Identity];
```

```
extend[x_] := Append[x, First[x]]

Show[main, Graphics[
    Table[Line[extend[Rest[DuffingOrbits]]], {t, 0, 7, 1}]],
  PlotRange ->{{-1.7, 1.7}, {-1.3, 1.3}},
  AspectRatio -> Automatic, Frame -> True, Axes -> None,
  DisplayFunction -> $DisplayFunction];
```

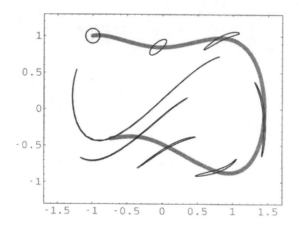

15.4 ■ Forced Attraction

If we vary the parameters a little (reduce the forcing amplitude to 0.25 and increase the damping coefficient to 0.25), then the Duffing orbits are much more well behaved: they end up in one of the gravity wells near $(-1, 0)$ or $(1, 0)$. The following example is typical: after some oscillations, it falls into the left well and stays there.

```
SecondOrderPlot[Duffing[1, 1, .25, .25], x[t],
    {t, 0, 60}, {x, -1.7, 1.7}, {x', -1.7, 1.7},
    InitialValues -> {1, 1.1}, FastPlotting -> True,
    MaxSteps -> 800];
```

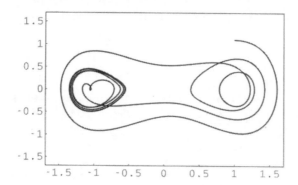

A plot of x vs. t shows the behavior a little more clearly.

```
SecondOrderPlot[Duffing[1, 1, 0.25, 0.25], x[t],
   {t, 0, 150}, PlotVariables -> {{x}}, InitialValues -> {1, 1.1},
   FastPlotting -> True, PlotRange -> All, MaxSteps -> 2000];
```

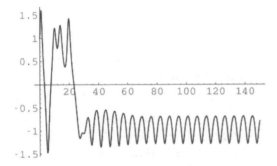

It is not an easy task to determine which well will attract a given initial value. Unlike the autonomous case, the phase plane is not simply partitioned by the orbits. So there is nothing for it but to look at thousands of initial conditions and simply keep track of where they end up. Fortunately, DensityPlot is fairly efficient. We proceed by defining AttractingWell to return $+1$ or -1 depending on where the orbit gets to. Then we use DensityPlot to plot this function. While some cases are stable for t out to 100 or more, other cases are not. Here we restrict ourselves to $t < 16\pi$. Thus we must be aware that some orbits will not have settled down yet. The color image is in color plate 23.

The code that follows computes 160,000 orbits and takes several hours.

```
tmax = 16 Pi;

AttractingWell[a_, b_] := Sign[(x[t] /. First[NDSolve[
   {Duffing[1, 1, .25, .25] , x[0] == a , x'[0] == b},
    x[t],  {t, 0, tmax}, MaxSteps -> 4000] )[[0, 2, -1, 3, 1]]]

DensityPlot[AttractingWell[a, b], {a, -2.5, 2.5}, {b, -2.5, 2.5},
   PlotPoints -> 400, Mesh->False,
   ColorFunction -> (If[# != 1, Blue, Yellow]&)];
```

15.5 ■ What To Do?

The sensitivity we have seen is not simply an anomaly of the numerical methods. It is in fact true that very small differences in the Duffing initial conditions can lead to very large differences in the orbits 100 seconds later. This is the often mentioned "butterfly effect," which refers tō the fact that a complicated system of differential equations, such as those that occur in weather prediction, might (note: "might", not "will") be sensitive to changes on the order of 10^{-10} or smaller. There are several different ways to deal with this problem.

Physical Reality

If it is not the Duffing equation we are so interested in, but the underlying physical phenomenon that it models, then we can take the point of view that, because we cannot measure initial conditions to more than one part in 10^{-10}, it is somewhat meaningless to ask for the true orbit.

Change the Question

The subsequent comments, which are related to those in the previous paragraph, were pointed out to us by Robert Corless, who has written on the usefulness of residuals [Cor].

Look at the residual (the `ResidualPlot` function, discussed in Chapter 2, can be used to generate its graph), denoted $r(t)$. Then the computed solution (an interpolating function) is a completely accurate solution to the perturbed differential equation, in which the $d \cos t$ term is replaced by $d \cos t + r(t)$. If $r(t)$ is less than, say, 10^{-10}, then we can't be certain that it is not present as noise in the physically real situation. This gives us a way of changing the question so that we are absolutely certain that the computed answer is the true answer to the new question.

The Shadow Knows!

In some sensitive cases it has been proven that computed orbits are actually true orbits, but for slightly perturbed initial conditions. While this phenomenon, called *shadowing*, has not been strictly proved for the Duffing equation, it is suspected to be true. From this point of view, one is seeing a true orbit, but not for the precise initial values presented.

Don't Give Up

All this might seem hopeless to the idealistic mathematician who wants the true solution (for some infinite precision initial conditions such as $(-1, 1)$). And in a sense it is hopeless! But in fact, *Mathematica*'s high-precision capabilities can give us some additional useful information. We must emphasize that our point here is not so much the solving of the Duffing equation, but the discussion of general techniques that can be used on any equation to help determine if the numerical algorithm is doing the right thing.

Our basic suggestion is to increase the precision goals. If we work in, say, 20-digit precision throughout and sharpen the local truncation error to, say, 10^{-10}, then we will maintain more precision at each step and get past the $t = 60$ barrier that stops machine precision cold. True, we won't get very much farther. But the idea is instructive to follow through. Instead of using `VisualDSolve`, it is simpler to use the built-in `NDSolve` command.

Warning. The commands that follow do not really work in version 2.2 of *Mathematica*, because the NDSolve function in that version does not behave as promised for high precision settings. Thus these commands and therefore this entire general approach—are meant for use with version 3.0. The way that the bad behavior in version 2 was discovered was by examining the number of steps the method was taking. We can see these numbers by counting the points in the interpolating function. As the precision goal increases, the number of steps should rise. This was not happening in version 2. To count the number of points, we define a simple utility that takes as input either IFcn or IFcn[t] where IFcn is an InterpolatingFunction, and returns the number of data points.

```
IntFuncLength[IFcn_[_]] := IntFuncLength[IFcn]
IntFuncLength[IFcn_] :=  Length[Last[IFcn]]
```

A quick test reveals that only 32 steps are needed by NDSolve to get a 6-significant digit solution to the simple equation $dx/dt = x(t)$.

```
test = x[t] /. First[
  NDSolve[{x'[t] == x[t], x[0] == 1}, x[t], {t, 0, 4}]];
IntFuncLength[test]

32
```

Here are two important points regarding increased precision:
- When using higher precision it is crucial that no floating-point reals be used in the differential equation. That is why we will use coefficients such as 3/10 and 15/100.
- When the working precision is set to a high value, say 24, then the precision goal, if it is set to Automatic, is automatically set to ten digits lower, or 14 (this is true in version 3.0 of *Mathematica*; in earlier versions it is more difficult to generate high-precision solutions).

```
Do[wpsol[wp] = x[t] /. First[NDSolve[{
  x"[t] - x[t] + x[t]^3 + 15/100 x'[t] == 3/10 Cos[t],
  x[0] == -1, x'[0] == 1}, x[t], {t, 0, 100},
  MaxSteps -> 4000, WorkingPrecision -> wp]],
    {wp, 16, 20, 2}];
```

We can now check the number of steps used. It is comforting to see this number grow at a reasonable rate.

```
Table[IntFuncLength[wpsol[i]], {i, 16, 20, 2}]

{1089,1595,2279}
```

Now we look at a single perturbed solution, using the precision-goal-8 example (working precision equals 18).

```
HighPrecPerturbed = x[t] /. First[
  NDSolve[{x"[t] - x[t] + x[t]^3 + 15/100 x'[t] == 3/10 Cos[t],
        x[0] == -1 + 10 ^ -8, x'[0] == 1 + 10 ^ -8},
        x[t], {t, 0, 100},
        MaxSteps -> 4000, WorkingPrecision -> 18]];
```

When we plot the two solutions for $x(t)$ we see that they are in agreement out to about $t = 90$, well past the $t = 60$ barrier for machine precision.

```
Plot[{wpsol[18], HighPrecPerturbed}, {t, 0, 100}]
```

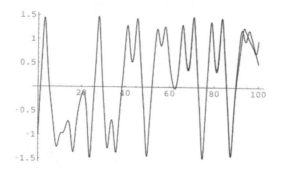

Using these ideas it is possible to get somewhat past $t = 100$. However, some authors have computed Poincaré sections of the Duffing equation to $t = 10,000$ and beyond. Rob Knapp and Stan Wagon [KW] have used Lyapunov exponents to estimate that 800 digits of precision would be required if the computation were to pass the present tests for accuracy. This is not in itself an insurmountable obstacle. But they also estimated that 10^{66} steps would be needed!

We conclude by presenting the orbit out to $t = 100$, generated using 24 digits of working precision and a precision goal of 14. We believe that this orbit, unlike previously published ones, is an accurate representation to within 1 part in 10^{-6} of the solution for the initial conditions $(-1, 1)$.

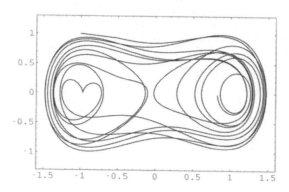

Chapter 16

The Tetrapods of Wada

Overview

This chapter deals with the fairly simple differential equation governing the motion of a damped pendulum subject to a periodic external force. It seems as if, no matter the initial condition, it will eventually settle into oscillatory behavior. But how many times will it rotate before falling into the terminal loop? This question is amazingly complex: the set of initial conditions leading to oscillation after a fixed number of rotations forms a remarkable region of the plane known as a Lake of Wada. In this chapter we show how to generate a striking image of such lakes. It is quite remarkable that such complexity arises from a fairly simple physical object. This result was proved by Hubbard ([Hub1]; see also [Hub]), using methods developed by Kennedy, Nusse, and Yorke [KY1, KY2, NY1, NY2, NY3]. To be more precise, Yorke and his co-workers found a pendulum with three Wada basins, and Hubbard extended their methods to an example with infinitely many Wada basins, and it is that example that we present here. This material is useful for the inherent beauty in the ultimate images, but also because it is a good example of the meaning and significance of the Poincaré map.

16.1 ◼ A Damped, Forced Pendulum

If the standard pendulum motion is enhanced by adding a damping term proportional to the velocity and a forcing term that feeds periodic energy into it, one gets the following system.

```
eqn = (x"[t] == -0.1 x'[t] - Sin[x[t]] + Cos[t]);
```

ToSystem converts the equation to a first-order system, which we call sys.

```
sys = ToSystem[eqn, x[t], t, AuxiliaryVariables -> y]
{x'[t] == y[t], y'[t] == Cos[t] - Sin[x[t]] - 0.1 y[t]}
```

If the pendulum is given an initial velocity, then the evidence is that it will eventually settle into periodic oscillation around its stable equilibrium—the straight-down position—but the number of times it spins (counting clockwise spin as negative!) before reaching that oscillation is difficult to predict. The following image shows x vs. t for eleven different initial velocities between 1.85 and 2.1 (for a color version of this image see color plate 4).

```
SystemSolutionPlot[sys, {x[t], y[t]}, {t, 0, 200},
    InitialValues -> Table[{0, y}, {y, 1.85, 2.1, 0.025}],
    PlotVariables -> {{x}}, MaxSteps -> 3000,
    PlotPoints -> 200, AxesOrigin -> {0, -35},
    PlotRange -> {-35, 23}];
```

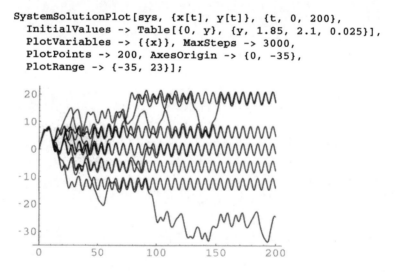

The curve that straddles the x-axis represents a solution that makes zero full spins (though it is possible that it spins several times in one direction before cancelling those revolutions with spins in the opposite direction).

Note that in the absence of a driving force, the behavior of the damped pendulum is very well-behaved; see the shaded nullcline image on page 103. And if there is no damping, then the forced pendulums behavior is chaotic (see [Str, exercise 12.5.4]). Here is an image to illustrate that.

```
SecondOrderPlot[x"[t] == -Sin[x[t]] + Cos[t], x[t],
   {t, 0, 24 Pi}, InitialValues -> {0, 2},
   MaxSteps -> 2000,
   PlotVariables -> {{x}}, PlotRange -> All];
```

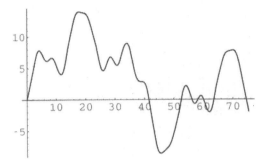

16.2 ■ Finding the Cycles

It greatly simplifies things to look at the Poincaré map, *P*, which views the orbits every 2π seconds. This yields images that are easy to interpret, and somewhat avoids the time dependence. For example, one way to determine the exact location of the cycles is to examine the phase portrait.

```
PhasePlot[sys, {x[t], y[t]},
   {t, 0, 30 Pi}, {x, -16, 4}, {y, -4, 4},
   InitialValues -> {0.3, 0.3}, MaxSteps -> 1000,
   ShowInitialValues -> True, AxesLabel -> {x, y}];
```

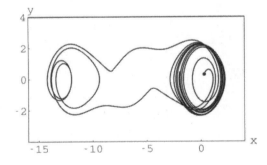

We can see that, after a side trip to the west, the orbit settles down to a cycle around the origin. But these images (unlike orbits for autonomous systems) tend to yield tangled messes. The Poincaré map reduces the study to points. We use *P* to denote the function that takes a point (*a*, *b*) to the position of the orbit that starts at (*a*, *b*) and runs for 2π seconds. Note that the system at hand is not autonomous but, because the forcing function is periodic with period 2π, there is no need to specify a time input.

We first define the Poincaré map to handle arbitrary time intervals, but with a default of 2π. We then pull out two special cases for convenience later: P2 is the 4π Poincaré map, while PInv is the inverse of the 2π map. Although we could use PoincareSection and its options to define P, it is just as easy, and a little faster, to use NDSolve directly.

```
pi = N[Pi];

P[{a_, b_}, int_:2 pi] := {x[t], y[t]} /. First[NDSolve[
    Join[sys, {x[0] == a, y[0] == b}], {x[t], y[t]},
        {t, 0, int}]] /. t -> N[int]

P2Inv[{a_, b_}] := P[{a, b}, -4 pi]

P2[{a_, b_}] := P[{a, b}, 4 pi]
```

The next image shows the location of the points obtained by iterating P. This corresponds to sampling the orbit every 2π seconds. Note how the points converge to a single value. That value is an attracting point for P, which we call a *sink* for the Poincaré map.

```
PhasePlot[sys, {x[t], y[t]},
    {t, 0, 30 Pi}, {x, -16, 4}, {y, -4, 4},
    InitialValues -> {0.3, 0.3}, MaxSteps -> 1000,
    AxesLabel -> {x, y}, GrayShading -> True,
    ParametricPlotFunction -> ColorParametricPlot,
    Epilog -> {AbsolutePointSize[5],
            Map[Point, NestList[P, {0.3, 0.3}, 15]]}];
```

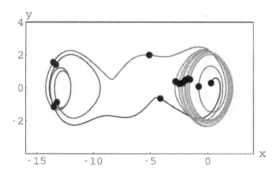

Now we may as well look only at the points generated by P.

```
Show[Graphics[{AbsolutePointSize[4],
    Map[Point, pOrbit = NestList[P, {0.3, 0.3}, 60]]}],
    PlotRange -> All, Frame -> True];
```

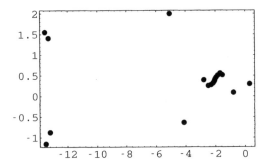

Looking at the last five points shows that convergence has been achieved.

```
Take[pOrbit, -5]
```

{{-2.04629, 0.392688}, {-2.04629, 0.392687}, {-2.04629, 0.392688},
    {-2.04629, 0.392688}, {-2.04629, 0.392688}}

Now we look at the final points of sequences obtained by iterating P twenty times on thirty random inputs.

```
Show[Graphics[{AbsolutePointSize[5],
   Map[Point, Map[Nest[P, #, 20]&, Table[
     {Random[Real, {-10, 10}], Random[Real, {-1, 1}]}, {30}]]]}],
   Frame -> True, PlotRange -> {{-23, 19}, {-2, 2}} ];
```

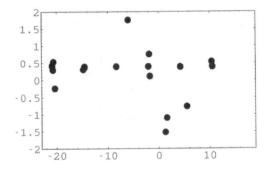

We see that, in several cases, 20 iterations were not sufficient for convergence. But most of the inputs led to a point on the line $y = 0.39$. We now define the main sink to be the one corresponding to zero spins.

```
sink[0] = Last[pOrbit]
```

{-2.04629, 0.392688}

We can verify the zero-spin nature of this sink by plotting the solution. The angle stays between -2.1 and 2.1 radians.

```
SystemSolutionPlot[sys, {x[t], y[t]}, {t, 0, 10 Pi},
   InitialValues -> sink[0], SaveSolution -> True,
   PlotVariables -> {{x}}];
```

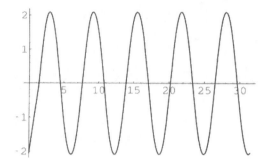

Of course, there are infinitely many sinks, the others differing from the main one by multiples of 2π.

```
sink[n_] := sink[0] + {2 n pi, 0}

Show[Graphics[{AbsolutePointSize[6],
   Table[Point[sink[n]], {n, -4, 2}]}], Frame -> True];
```

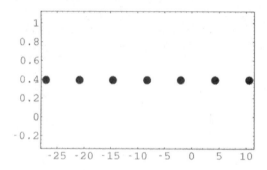

16.3 ■ Surprising Periodicity

Our computations so far might lead us to think that all initial values lead the pendulum to one of infinitely many attracting cycles (or, in terms of P, lead to an orbit that converges to a sink). But this is not true! Certainly if the starting value is near a sink, then the sequence will converge to that sink. But as the starting value moves away from the sink, the convergence becomes slower and eventually one enters a region of convergence to another sink. Thus one should ask: what happens to initial values that are right on the border between these two behaviors?

A first observation is that such a boundary point must stay on the boundary. (Sketch of proof: If (a, b) led to a sink then, after some fixed number, n say, of iterations, it would be within 0.01 of the the sink. Then, by continuity, there would be a disk around (*a*, *b*) such that points in the disk, after *n* iterations, get taken to within 0.02 of the sink, close enough to guarantee convergence.) Let us first turn to the job of finding a point on the border. If (*a*, *b*) is on the border, then it seems reasonable that the distance between the tenth *P*-iterate of a point just below (*a*, *b*) and the tenth *P*-iterate of a point just above (*a*, *b*) will be great. We can test this hypothesis by examining a plot of such distances as we head north from (−2.05, 0), a point close to the main sink.

```
dist[{p_, q_}] := N[Sqrt[(p - q) . (p - q)]];

ListPlot[Table[{b, Abs[dist[{sink[0], Nest[P, {-2.05, b}, 10]}] -
            dist[{sink[0], Nest[P, {-2.05, b + 0.001}, 10]}]]},
    {b, 0.1, 0.9, 0.02}], PlotJoined -> True,
    Frame -> True, PlotRange -> {-1, 15}];
```

The distance difference does indeed jump near 0.85. We can get this critical height more accurately by finding the maximum of the function just plotted.

```
basinPt = {-2.05, b} /. Last[FindMinimum[-Abs[
        dist[{sink[0], Nest[P, {-2.05, b}, 10]}] -
        dist[{sink[0], Nest[P, {-2.05, b + 0.001}, 10]}]]],
    {b, {0.8, 0.9}}]]

{-2.05, 0.855731}
```

Thus we have a point that is very near to the basin boundary. Now, using the investigations by Hubbard [Hub1] as a guide, we look for a point that is periodic of period 4π. Of course, every sink, being 2π-periodic, is also 4π periodic. But it turns out that there are points that are 4π-periodic and not 2π-periodic. Indeed, there might even be other types of periodic behavior for this example; Hubbard [Hub1] observes that this is an unresolved question.

We can ask *Mathematica* to search for the 4π-periodic point by using FindRoot to attack the two-dimensional root-finding problem: P2[{x, y}] == {x, y}. Typical initial values would lead to one of the sinks. But our basin boundary point does not!

We feed two initial values for *a* and two for *b* to FindRoot because otherwise it would try to use Newton's method, which would get stuck because of the lack of a symbolic derivative for P2.

```
e = 0.001;

Find4PiPeriodic[{a_, b_}] := {xx, yy} /.
  FindRoot[P2[{xx, yy}] == {xx, yy},
    {xx, a + {-e, e}}, {yy, b + {-e, e}}, AccuracyGoal -> 12]

p = Find4PiPeriodic[basinPt]
{-2.00037, 0.865221}
```

Let's look at the solution corresponding to this interesting point.

```
SystemSolutionPlot[sys, {x[t], y[t]}, {t, 0, 20 Pi},
  InitialValues -> p, PlotVariables -> {{x}},
  MaxSteps -> 1000];
```

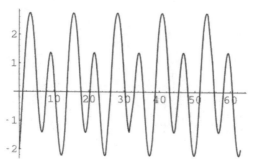

The 4π periodicity is evident. It is as if the solution can't make up its mind which sink to converge to, so it bounces around forever without converging to any of the sinks. Here is the phase plane view.

```
PhasePlot[sys, {x[t], y[t]}, {t, 0, 8 Pi},
  {x, -3, 3}, {y, -3, 3}, ShowInitialValues -> True,
  InitialValues -> p];
```

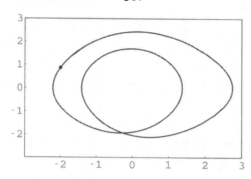

It is evident from the plot of x vs. t that $P(p)$ is also a 4π-periodic point. We can get this other periodic point more accurately by again using Find4PiPeriodic.

```
Find4PiPeriodic[P[p]]
```

```
{-1.39637, -0.233485}
```

Similar work leads to two more periodic points. Here is a full-precision list of the four periodic points, followed by phase plane images of the two 4π-periodic solutions with the four periodic points superimposed. Animate these two images to see how similar they are.

```
periodic[1] = {-2.742031575604344, 0.07271751909094519};
periodic[2] = {-1.229428020159861, 0.7350878694723261};
periodic[3] = {-2.000374755607681, 0.865221158707783};
periodic[4] = {-1.396365149945631, -0.2334852341993026};

periodics = Map[periodic, Range[4]];

PhasePlot[sys, {x[t], y[t]}, {t, 0, 4 Pi},
    {x, -3, 3}, {y, -2.5, 2.5}, InitialValues -> periodic[1],
    Epilog -> {AbsolutePointSize[5], Map[Point, periodics]}];
```

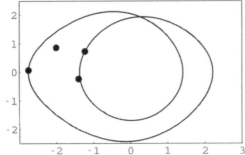

```
PhasePlot[sys, {x[t], y[t]}, {t, 0, 4 Pi},
    {x, -3, 3}, {y, -2.5, 2.5}, InitialValues -> periodic[3],
    Epilog -> {AbsolutePointSize[5], Map[Point, periodics]}];
```

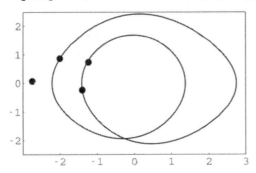

16.4 █ The Tetrapod

Our goal now is to generate a view of the basins of attraction to the sinks. This material is somewhat more advanced than the rest of this book, and we make no attempt to prove or justify many of the claims (the interested reader should consult [Hub1] and the references therein). Rather, we will present the main steps of an algorithm to generate an eye-opening image of the basins for the forced, damped pendulum.

To put this problem in context, recall that Chapter 15 had two examples of basin drawings for the Duffing equation. In the first case (page 208) it sufficed to draw a single separatrix, obtained as a solution to a differential equation. In the second case (color plate 23) we took a time-consuming pixel-by-pixel route. For the case at hand we will get the basins by drawing curves, but it is not so simple as generating the solution to a single differential equation.

A natural place to start is to try to understand the behavior of the 4π-periodic Poincaré map (henceforth P_2, because it is two iterations of P, the standard 2π map) near the periodic points. This behavior is saddle-like, as can be seen by linearization as follows. Here we are approximating the derivative of P_2 with a centered finite difference based on $\Delta x = 0.001$.

```
deltax = 0.001;
saddle = periodic[1];
Eigensystem[Transpose[
   {P2[saddle + {deltax, 0}] - P2[saddle - {deltax, 0}],
    P2[saddle + {0, deltax}] - P2[saddle - {0, deltax}]} /
      (2 deltax)]]

{{2.92704, 0.0974444}, {{0.901938, 0.431865},
   {-0.670284, 0.742104}}}
```

This tells us that, near the first periodic point, P_2 expands by a factor near 3 along a line in the ENE direction and contracts by about 1/10 in the NW direction. Warning: We are linearizing a map here; this is different than linearizing a differential equation, where negative eigenvalues correspond to convergence and positive ones to divergence.

Now we use *Mathematica*'s Arrows package and a unit-vector function to help us draw some equal-length arrows indicating the direction P_2 takes near periodic[1]. Because x and y are used in the definition of the basic system, it is important to use other variables when defining the table, since that calls P2, which in turn calls sys.

```
evs = Last[%];
Needs["Graphics`Arrow`"];

unit[v_] := v / Sqrt[v.v]
```

```
Show[Graphics[{{AbsoluteThickness[2.5],
   GrayLevel[0.25],
   Line[{saddle + evs[[1]], saddle - evs[[1]]}],
   GrayLevel[0.75],
   Line[{saddle + evs[[2]], saddle - evs[[2]]}]},
   {AbsoluteThickness[0.5], Table[
     Arrow[{xx, yy}, {xx, yy} + unit[(P2[{xx, yy}] - {xx, yy})]/70,
       HeadLength -> 0.025],
     {xx, saddle[[1]] - 0.11, saddle[[1]] + 0.1, 0.02},
     {yy, saddle[[2]] - 0.11, saddle[[2]] + 0.1, 0.02}]},
   AbsolutePointSize[8], Point[saddle]}], AspectRatio -> Automatic,
   Frame -> True, PlotRange -> {{-2.86, -2.64}, {-0.04, 0.18}}];
```

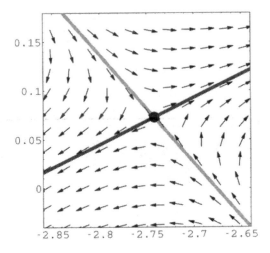

In the preceding diagram the dark gray line is the asymptotic direction of the attracting curve for P_2 (in dynamical systems terminology, this is part of the *unstable manifold*; the other eigenspace gives the asymptotic direction of the *stable manifold*). The separatrix that we want is the curve consisting of points that P_2 takes to the saddle point. Thus we can now see the way to our goal. We can start at just about any point near the saddle, draw a line from the saddle to that point, compute the image of that line under the *inverse* Poincaré map for period 4π (defined as P2Inv on page 224). That image will be very close to the curve we want, and we can then repeat the process using the piece of the curve we now have. Because P2Inv takes large jumps, repeating this three or four times should suffice to get a large chunk of the curve.

The downside is that these large jumps make it very difficult to get smooth output. Thus our algorithm must constantly leapfrog from range to domain to compute values at enough points to generate a smooth curve. Since we know the eigenvectors, we may as well start with a point on the right eigenspace (but, in fact, this is not essential; we can start at almost any point near the saddle). We always want the eigenvector with large eigenvalue, and that will be the first in the list.

The code for implementing these ideas is relegated to the next section. For now, let us look at some of the results. Getting all the data to make a nice picture takes a couple of hours, so the disk that accompanies this book contains a package, `Wada.m`, that automatically loads the data into four variables: `basin[1]`, `basin[2]`, `basin[3]`, `basin[4]`. Each variable will contain points on the separatrix (in both directions) through one of the periodic points. Here is how to load and use this information.

```
Needs["Wada`"]

Length[basin[1]]

1067
```

Let's look at a small portion of this curve: the set consisting of 80 points in both directions from the periodic point.

```
middle = Round[Length[basin[1]]/2];
Show[Graphics[{
  AbsolutePointSize[5], Point[periodic[1]],
  Line[Take[basin[1], {middle - 80, middle + 80}]]}],
  Frame -> True];
```

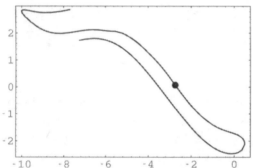

The following code generates an amusing movie of the entire first separatrix. We include here only a single image showing the whole curve.

```
step = 15;
Do[Show[Graphics[{AbsolutePointSize[5], Point[periodic[1]],
  {AbsoluteThickness[2],
   Line[Take[basin[1], {middle + (i-step), middle + i}]],
   Line[Take[basin[1], {middle - i, middle - (i-step)}]]},
  {AbsoluteThickness[0.5],
   Line[Take[basin[1], {middle - (i - step), middle + (i-step)}]]}}],
  Frame -> True, PlotRange -> {{-14, 34}, {-6, 4.2}}],
 {i, step, 533, step}]
```

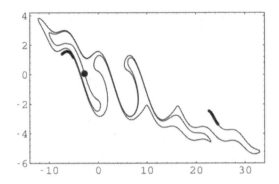

When we look at all four separatrices at once, we begin to glimpse the complexity of the situation.

```
Show[Graphics[{AbsoluteThickness[0.4],
    Map[Line, Map[basin, Range[4]]]}, Frame -> True]];
```

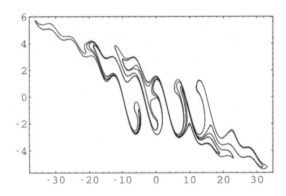

The complexity of this image is difficult to grasp. Let's backtrack and look at a much smaller piece of each curve.

```
smallBasin[1] = Take[basin[1], {middle - 36, -middle + 47}];
smallBasin[2] = Take[basin[2], {middle - 17, -middle + 12}];
smallBasin[3] = Take[basin[3], {middle - 17, -middle + 39}];
smallBasin[4] = Take[basin[4], {middle -  5, -middle + 22}];

tetrapod = Show[Graphics[{
    AbsoluteThickness[0.5], Line /@ smallBasin /@ Range[4],
    Disk[sink[0], 0.1],
    {GrayLevel[0.6],
      Disk[periodic[1], 0.1], Disk[periodic[2], 0.1]},
    {White,
      Disk[periodic[3], 0.1], Disk[periodic[4], 0.1]},
    Table[Circle[periodic[i], 0.1], {i, 1, 4}]}],
  Frame -> True, AspectRatio -> Automatic,
  FrameTicks -> {Automatic, Automatic, None, None}];
```

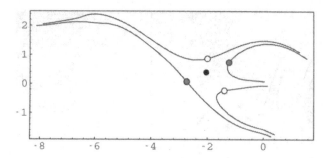

The main sink, $(-2.05, 0.39)$, is the point in the center and the other four points are the periodic points. Here is what happens, where we keep with the Poincaré map viewpoint, which is so much more convenient for this, and similar, examples.

- The *P*-orbit of any point inside the 4-armed region—the tetrapod—converges to the sink.
- The four periodic points just cycle among themselves in pairs.
- Points on the bounding curves yield orbits that bounce from curve to curve and converge to a periodic pair. Looking at the 4π-Poincaré map instead, such points have orbits that converge to one of the periodic points.

The following image shows the result of two initial conditions: the one at the eye of the tetrapod heads in to the sink, while the one at the bottom right heads to the white periodic points.

```
Show[tetrapod, Graphics[{AbsolutePointSize[3],
    Map[Point, NestList[P, {-6.07, 2.2}, 10]],
    GrayLevel[0.3],
    Map[Point, NestList[P, smallBasin[4][[-1]], 10]]}]];
```

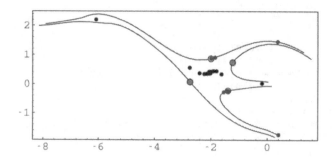

Now that we have the basic tetrapod and understand its role as a basin of attraction, we must try to understand what happens when we allow the tetrapod to grow its four tendrils without bound. And also we should look at the various translations of the creature, since they correspond to basins of attraction for the other sinks. Color plate 24 contains an image showing several of these creatures; the following code generates the color image.

First we truncate the curves to yield a more manageable polygon object; joining them in the right order gets us the polygon, which we call tetrapod.

```
smallBasin[1] = Take[basin[1], {middle -  50, -middle +  70}];
smallBasin[2] = Take[basin[2], {middle -  44, -middle + 117}];
smallBasin[3] = Take[basin[3], {middle - 107, -middle +  57}];
smallBasin[4] = Take[basin[4], {middle -  45, -middle +  31}];

tetrapod = Polygon[Join[smallBasin[1], Reverse[smallBasin[3]],
                   Reverse[smallBasin[2]], smallBasin[4]]];

Show[Graphics[{AbsolutePointSize[4], tetrapod}]];
```

Now we show the tetrapod and treat the rest of the curves as simple Line objects. The following code does the proper translation and coloring to generate a striking image (see color plate 24).

```
Needs["Graphics`Colors`"]

{xmin, xmax, ymin,  ymax} = {-14, 12, -3.3, 4.3};
tr[l_, n_] := l /. {x_?NumberQ, y_?NumberQ} :> {x + n 2 pi, y}

Show[Graphics[{AbsolutePointSize[4],
   bb = {Blue, tetrapod,
      AbsoluteThickness[1], Line /@ basin /@ Range[4]},
   tr[bb /. {Blue -> Red}, 1], tr[bb /. {Blue -> Yellow}, -1],
   tr[bb /. {Blue -> LightBlue}, 2], tr[bb /. {Blue -> Magenta}, -2],
   tr[bb /. {Blue -> Green}, 3],
   {Black, Point /@ sink /@ Range[-2, 3]},
   {Black, Table[Map[Point, tr[periodics, j]], {j, -2, 3}]}}],
   Background -> GrayLevel[0.4],
   Epilog -> {AbsoluteThickness[3], Black,
      Line[{{xmin, ymin}, {xmax, ymin}, {xmax, ymax},
            {xmin, ymax}, {xmin, ymin}}]},
   PlotRange -> {{xmin, xmax}, {ymin, ymax}}];
```

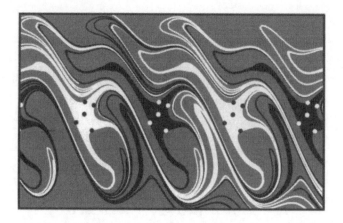

The Lakes of Wada are a mathematical construction named after a Japanese mathematician who, in the early 20th century, showed how to construct three disjoint, connected, open subsets of the plane so that any point on the boundary of one of them was also on the boundary of both of the two others. Loosely speaking, imagine people whose motion is restricted to one of three open sets, called *A*, *B*, and *C*. Then any *A*-person (no matter how small) can pick any point at the border of *A*, and go to a point in *A* close enough to that border point so that he or she can shake hands with a similarly small person who lives in region *B* or *C*. Now, the amazing thing is that the basins of attraction for the sinks form infinitely many Lakes of Wada! This means that they are so incredibly intertwined that any point in the boundary of one of the basins is in the boundary of all of the others! It is quite remarkable that such complexity arises from a fairly simple physical object. This result was proved by Hubbard [Hub1], using methods developed by Yorke and his co-workers.

As noted, we will not go into the proofs of these properties. However, there is more to say about this picture. The ultimate goal, of course, is to understand the possible behaviors of the pendulum.

1. If the initial condition lies in a basin, then the pendulum ends up in one of the oscillatory loops.
2. If the initial condition lies exactly on the curves forming the separatrices, then the solution converges to one of the periodic pairs.
3. But there is more! The separatrices form only what is called the "accessible basin boundary," those points on the boundary that can be reached from a sink by a path contained in the basin (*i.e.*, a boat ride that stays in the lake). In fact, the topological boundary of the basin is more complicated, and has points that are not on the curves defining the borders (recall the formal definition: the boundary of an open set is the collection of its limit points, excluding the open set itself).

For a simpler example of this sort of thing, consider the curve that is the graph of $\sin(1/x)$ between 0 and 1. Fatten it a bit and consider it as a union of polygons.

Then the entire union is a connected open set. Its accessible boundary consists of the points obtainable by moving up or down from a point on the graph to the edge of the region. But the segment on the *y*-axis between −1 and +1 is also part of the boundary of the region; it is not accessible from points inside the region.

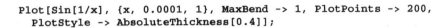

```
Plot[Sin[1/x], {x, 0.0001, 1}, MaxBend -> 1, PlotPoints -> 200,
     PlotStyle -> AbsoluteThickness[0.4]];
```

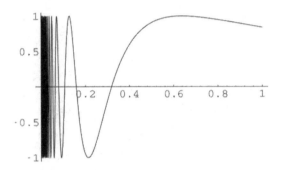

The Wada basins are much more complicated than this, involving limit points of limit points, and so on. One thing that is not completely clear is whether the basins are dense in the plane. Could there be an entire region that they avoid? And even if they are dense, does the set of points not in a basin have measure 0? As observed by Hubbard, a positive answer to this last question would mean that, with probability one, the pendulum will end up at a sink.

16.5 ■ The Curve-Drawing Algorithm

The code that follows contains everything needed to generate the data for the curves and store them in basin[1], basin[2], basin[3], basin[4]. The auxiliary functions and initializations are either self-explanatory or have been explained earlier in the chapter. The tricky part is the algorithm that draws the curve, which is implemented as Separatrix. It uses a leapfrogging technique (thanks to John Hubbard for the big picture, and to Joan Hutchinson for helping with some tricky programming).

Recall that the main idea is to start with a point in the direction of an eigenspace, and apply P2Inv to it. Separatrix builds an array of *pairs* of points, called sepPts. The first point in a pair is one of the points on the final curve. The second point is its predecessor under the P2Inv map. Now at each stage we have a target, which is P2Inv of one of the points on the already defined curve. Typically the target will be quite far away from the current end of the curve. We use linear interpolation on a certain interval (this is carried out via the sc construction in the While loop) to find a point within a certain tolerance of the current end. Eventually, we reach the current target, at which point we get a new target by taking P2Inv of the point in the curve-list right after the one used to get the current target.

In short, we leapfrog back and forth between points on the curve and their P_2 images and inverse images so that the end-result is a smooth curve.

There are a few options that are explained in their usage messages. The Type option has a default of Stable which causes the routine to draw the stable manifold, which is the separatrix we want. But to get the unstable manifold we have simply to change P2Inv to P2 at three places in the program. So an option is included for users who wish to view the unstable manifolds.

```
Needs["Graphics`Colors`"]

Off[General::spell1];

pi = N[Pi];

sys = {x'[t] == y[t],
        y'[t] == -0.1 y[t] - Sin[x[t]] + Cos[t]};

(* The squared-distance function, compiled for speed *)
func = Compile[{a, b, c, d}, (a-c)^2 + (b-d)^2];
distSq[{{a_, b_}, {c_, d_}}] := func[a, b, c, d]

(* The Poincare map, followed by two special cases *)
P[{a_, b_}, int_:2 Pi] := {x[t], y[t]} /.
    First[NDSolve[Join[sys, {x[0] == a, y[0] == b}],
      {x[t], y[t]}, {t, 0, int}]] /. t -> N[int]
P2Inv[{a_, b_}] := P[{a, b}, -4 pi]
P2[{a_, b_}] :=  P[{a, b}, 4 pi]
eigen[pt_] := First[Eigenvectors[Transpose[
  {P2[pt + {0.001, 0}], P2[pt + {0, 0.001}]}]]]

(* The five special points, and their translates *)
periodic[1] = {-2.742031575604344, 0.07271751909094519};
periodic[2] = {-1.229428020159861, 0.7350878694723261};
periodic[3] = {-2.000374755607681, 0.865221158707783};
periodic[4] = {-1.39636514994563, -0.2334852341993176};
periodics = Map[periodic, Range[4]];

sink[0] = {-2.046289187487275, 0.3926876640601063};
sink[n_] := sink[0] + {2 n pi, 0};

(* A general translation function *)
tr[l_, n_] := l /. {x_?NumberQ, y_?NumberQ} :> {x + n 2 pi, y}

(* A function that thins a set by taking every nth element *)
thin[l_, n_] := l[[Range[1, Length[l], n]]]

(* The separatrix-constructing function *)
Tolerance::usage = "Tolerance is an option to Separatrix that
controls the maximum distance between successive points on the
curve. Too small a value can lead to errors in the curve.";
```

```
Direction::usage = "Direction is an option to Separatrix that
specifies the starting direction in terms of the eigenvectors
at the periodic point.";

ImageEnRouteSteps::usage = "ImageEnRouteSteps is an option to
Separatrix that asks for an image to be shown as the computation
proceeds. If set to m, then an image will be shown after every
m steps.";

Type::usage = "Type is an option to Separatrix that specifies
whether the Stable or Unstable manifold is to be drawn. The
curves separating the basins are given by the stable case.";

Options[Separatrix] =
 {Tolerance -> 0.25, Direction -> 1, ImageEnRouteSteps -> 10^9,
  Type -> Stable};

Separatrix[pt_, n_, opts___] := (
  {dir, tol, imStep, typ} =
    {Direction, Tolerance, ImageEnRouteSteps, Type} /.
    {opts} /. Options[Separatrix];
  sepPts = Array[Null &, n];
  poin = If[typ === Stable, P2Inv, P2];
  tolSq = tol^2;
  sepPts[[1]] = {pt, pt};
  sepPts[[2]] = Table[pt + 0.001 * dir * eigen[pt], {2}];
  marker = 2;
  target = poin[sepPts[[2, 1]]];
  Do[If[distSq[{sepPts[[ii - 1, 1]], target}] < tolSq, (
        sepPts[[ii]] = {target, sepPts[[marker, 1]]};
        marker++; target = poin[sepPts[[marker, 1]]]),
      (* else *) temp = sepPts[[ii - 1, 1]];
      sc = 0.4;
      While[distSq[{sepPts[[ii - 1, 1]], Ptemp = poin[temp]}] > tolSq,
        sc *= 0.2;
        temp = sc * sepPts[[marker, 1]] + (1 - sc)*(sepPts[[ii - 1, 2]])];
        sepPts[[ii]] = {Ptemp, temp}];

    If[Mod[ii, imStep] == 0, Show[Graphics[{AbsolutePointSize[5],
    Red, Point[target],
    Blue, Point[First[sepPts[[ii]]]]],
      Line[ First /@ Take[sepPts, ii-1]]},
    Frame -> True, PlotLabel->ii, PlotRange -> All]],
      {ii, 3, n}];
  First /@ sepPts)

n = 1600;
(* n = 350 is the smallest reasonable value to use. Good results are
   obtained with anything between 350 and 1600

   The package Wada.m causes the data to be loaded into the basin
   variables, thus saving hours of computing time.
*)
```

```
(* generate the eight basin curves in four groups of two *)
Do[Print[i]; (* to monitor things *)
   basin[i] = Join[
      Reverse[Separatrix[periodic[i], n, Direction -> -1,
                      Tolerance -> 0.15]],
             Separatrix[periodic[i], n, Tolerance -> 0.15]], {i, 1, 4}

(* thin out the basins, since not all the points are essential *)
Do[basin[i] = thin[basin[i], 3], {i, 1, 4}]

(* to save the data you have generated to a file, you can use,
   for example, Map[basin, Range[1, 4]] >> Wada.dat
*)
```

Chapter 17

I Tossed a Book into the Air . . .

Overview

In this final chapter we examine a very accessible, but yet very complicated modeling problem: the rotational motion of a rigid body. In particular, we wish to investigate the apparent instability when an object such as a book is flipped in the air with the cover facing the thrower (*i.e.*, it spins around its medium-length axis).

In fact, the reader should right now get a book, wrap an elastic band around it to keep it closed, and start spinning it. A toss around the smallest of the axes of symmetry (just grab a corner and flick your wrist) yields boring spin: it is very easy to throw and catch the book. And a toss about the longest axis (grab the spine with both hands and spin) likewise yields predictable behavior. But if you hold the book along its bottom edge so that the title faces you, you will find that it is extremely difficult to spin it and catch it after one full turn so that the title faces you. The book, like a talented diver or gymnast, does an extra sideways flip, and lands with its back cover up!

Flipping a tennis racket into the air with the head facing up also makes a nice demonstration. If you flip it by the handle and try to catch the handle after one full rotation, it seems to always return upside down.

The discussion will contain several equations that we will not attempt to derive. A complete understanding would require a fair amount of physics background. Our goal is to show how the equations can be translated into enlightening animations that illustrate some of the beautiful classical theorems that describe the motion of a rigid body. For more detailed discussions, we recommend three references with very differing viewpoints; [Arn], [Mur] and [BCB].

This chapter is in part based on work done by Tamas Nemeth, a student in our differential course at Macalester College; he carried out some of these investigations as his final project.

17.1 ■ Euler's Equations

The motion of a rigid body can be uniquely written as the composition of a rotation about an axis through the center of mass and a translation of the body in a stationary coordinate system. Thus the motion decomposes into a rotational motion and a translational motion. We will consider only the rotational motion, because the translational motion is the same as the motion of a point mass, which is very simple to analyze and adds nothing to our understanding of the tumbling. Thus we wish to view the motion of a book rotating in space as one might see it aboard a space ship.

We denote by $\mathbf{k}$ the stationary coordinate system, with the usual x, y, and z axes, and whose origin, $\mathbf{0}$, represents the position of the book's center of mass. Coordinates in $\mathbf{k}$ represent the actual motion of the book (keeping in mind that we are taking the view of an observer on a space ship: the book is given an initial spin about an axis through its center of mass and so its center of mass remains fixed). And we are assuming that the book is really a box of uniform density; a real book has a spine that adds weight to one side. Our main goal, to be realized in section 17.3, is to generate a movie of the rotating book, which will require working in $\mathbf{k}$. A book has three axes of symmetry through its face centers, and it is clear that any initial spin whose axis *exactly* coincides with one of these three will just yield continuing spin about that axis. The interesting motion arises when the initial axis is slightly offset from one of the three main axes. Here is a movie that shows the tumbling book in the case of an initial spin axis offset a little from the long axis (the SpinTheBook function is defined in the initialization group at the beginning of the electronic version of this chapter, and will be discussed in section 17.3).

```
SpinTheBook[{0, 1, 10}, 2.5, ViewPoint -> {1.2, 1.4, 1.2},
    PlotRange -> {{-5, 5}, {-7, 7}, {-8, 8}}]
```

It is extremely useful to simplify still further and imagine what the tumbling book looks like to a very small bug who is on the book, not cognizant of the book's motion in space. As an analogy consider a human on the ever-spinning earth. The axis of spin actually changes a little bit over time (this is called *precession*). If we ignore the movement of the earth through space and the actual spinning of the earth, but look only at the spin-axis, then it would revolve around its average value. Thus we could imagine viewing a stationary sphere with a line through the rotation axis, and having that line move along the sphere as it in fact does. The analogy for the book is to fix the book in a horizontal position, say, and watch the rotation axis move. For example, the motion of the book that starts with a spin around an axis that is only slightly perturbed from the longest axis would be described by drawing the initial spin-axis and then having it wobble around the ideal symmetric axis. The code that follows generates a movie that shows the motion of this axis. The printed image is a composite from the animation.

```
SpinTheBook[{0, 1, 10}, 2.5, RealWorld -> False,
   RotationAxis -> True, Polhode -> True,
   ViewPoint -> {1.2, 1.4, 1.2},
   PlotRange -> {{-1, 1}, {-3,  3}, {-5, 12}}]
```

These ideas lead to the notion of a moving frame of reference (or inertial frame or coordinate system), which we call $\mathbf{K}$. We will do most of our work in $\mathbf{K}$; the tricky part (section 17.3) will be the transformation back to the stationary frame $\mathbf{k}$ to generate images of the actual rotation of the book. Let e_1, e_2, and e_3 denote unit vectors in the three coordinate axes of $\mathbf{K}$ and assume that the book lies in $\mathbf{K}$ so that its axes of symmetry are aligned with the e_i, its shortest axis corresponding to e_1, and the longest axis corresponding to e_3. Note that, because of the way we generate the book-images, e_1 will not correspond to the usual horizontal orientation of the x-axis on the screen.

If $\mathbf{q}$ denotes a position vector in the coordinate system $\mathbf{k}$ ($\mathbf{q}$, which represents a point on the book, is really $\mathbf{q}_t$, since it will change in time), then $\mathbf{Q}$ will be the corresponding position in the rotating coordinate system $\mathbf{K}$. These are related by a function $B : \mathbf{K} \to \mathbf{k}$ to be discussed in more detail in section 17.3.

The two key concepts we need to describe the motion of the body are angular velocity and angular momentum. The definitions are not as straightforward as velocity and momentum of a point mass. Recall that the instantaneous angular velocity vector, ω, of a rotating body is the vector whose direction is the axis of rotation and whose length is the angular speed in radians per second (with orientation given by the right-hand rule). A basic result is that $\mathbf{q}' = \omega \times \mathbf{q}$, where $\mathbf{a} \times \mathbf{b}$ denotes the cross product of the vectors $\mathbf{a}$ and $\mathbf{b}$ and $\mathbf{q}$ is the position vector. The corresponding vector, Ω, in the moving frame is given by $\Omega = B^{-1}\omega$ and is called the vector of angular velocity in the body. Our first goal will be to find and solve a system of differential equations for Ω, thus getting the bug's-eye view of the book's motion. First we define the three unknowns, which are the three components of Ω: $\omega_1(t)$, $\omega_2(t)$, $\omega_3(t)$.

```
OMEGA = Map[#[t] &, Array[w, 3]]
{w[1][t], w[2][t], w[3][t]}
```

For a rotating body, the analog of momentum for a point mass is the concept of angular momentum. We will use $\mathbf{M}$ for angular momentum in the moving frame $\mathbf{K}$. The symmetries of our book, which is placed in $\mathbf{K}$ so that its axes of symmetry correspond to the coordinate axes, allow for some nice simplifications (for less symmetrical objects one would call on the principal axes theorem of linear algebra). It turns out that $\mathbf{M}$, the angular momentum in $\mathbf{K}$, can be expressed as $(I_1\omega_1, I_2\omega_2, I_3\omega_3)$, where I_i is the moment of inertia with respect to the axis $\mathbf{e}_i$. For a system of point masses, $m_1, m_2, \ldots$, the moment of inertia with respect to an axis $\mathbf{e}$ is calculated as $I_\mathbf{e} = \sum m_j r_j^2$, where r_j is the distance from the point to the axis. In general, one would use an integral. In our application, the three moments of inertia are determined by the shape of the book: the largest moment would be for spin around the smallest axis, essentially because the corners of the book are far away from the axis. The book we use in our images has dimensions $4 \times 2.5 \times 0.5$, and so it is easy to compute the integrals for the moments of inertia, which turn out to be, up to a scalar multiple, 420, 307, and 123.

We are now ready to present Euler's equations, which describe how the angular velocities and momenta change over time. His result is simply:

$$\mathbf{M}' = \mathbf{M} \times \Omega$$

Using coordinates in $\mathbf{K}$, we can view this result as a system of three ordinary differential equations for Ω. We first define the abstract moments of inertia, as well as a rule to set the particular values.

```
Inertia = Array[i, 3]
{i[1], i[2], i[3]}
```

```
Ivals = {i[1] -> 420, i[2] -> 307, i[3] -> 123};
```

And here is the implementation of Euler's equations. We first define our own cross-product function.

```
CrossProduct[{u1_, u2_, u3_}, {v1_, v2_, v3_}] :=
  {u2 v3 - u3 v2, u3 v1 - u1 v3, u1 v2 - u2 v1}

TableForm[OMEGAeqns = Simplify[First[Solve[
  Thread[D[Inertia * OMEGA, t] ==
    CrossProduct[Inertia * OMEGA, OMEGA]], D[OMEGA, t]]]] /.
  Rule -> Equal]
```

$$(w[1])'[t] == \frac{(i[2] - i[3])\ w[2][t]\ w[3][t]}{i[1]}$$

$$(w[2])'[t] == \frac{(-i[1] + i[3])\ w[1][t]\ w[3][t]}{i[2]}$$

$$(w[3])'[t] == \frac{(i[1] - i[2])\ w[1][t]\ w[2][t]}{i[3]}$$

These equations are fairly simple, and *Mathematica* has no trouble solving them numerically. In fact, using an appropriate change of variables, Euler's equations can be solved in closed form using elliptic integrals. For a nice implementation of this point of view see the article by Hugh Murrell [Mur].

To view an orbit we first define a rule to set values for the three principal moments of inertia. An initial velocity of (0, 1, 10) corresponds to an initial spin around the longest axis, with a slight perturbation toward the medium axis. This leads to a fairly predictable wobble. We add a line along the long axis for reference.

```
Needs["VisualDSolve`"];

Show[PhasePlot[OMEGAeqns /. Ivals, Evaluate[OMEGA],
      {t, 0, 0.9}, {w[1], -4, 4}, {w[2], -4, 4}, {w[3], 0, 11},
      InitialValues -> {0, 1, 10}, BoxRatios -> {8, 8, 11},
      DisplayFunction -> Identity, Boxed -> True],
    Graphics3D[{GrayLevel[0.5], AbsoluteThickness[2],
      Line[{{0, 0, 0}, {0, 0, 10}}]}],
  DisplayFunction -> $DisplayFunction,
  ViewPoint -> {1, -2, 1.6}];
```

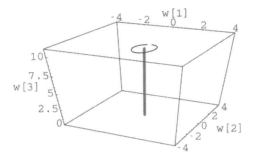

Using an initial spin of (10, 1, 0) yields a similar wobbly trajectory. But a choice of (0, 10, 1) corresponds to a perturbation around the medium-length axis. Since this turns out to be an unstable axis, the result should be interesting.

```
Show[
   PhasePlot[OMEGAeqns /. Ivals, Evaluate[OMEGA], {t, 0, 2.5},
      {w[1], -8, 8}, {w[2], -11, 13}, {w[3], 0, 11},
      InitialValues -> {0, 10, 1}, PlotPoints -> 200,
      SolutionName -> "OMEGAsoln",
      ShowInitialValues -> True, BoxRatios -> {14, 24, 11},
      Boxed -> True, DisplayFunction -> Identity],
   Graphics3D[{GrayLevel[0.5], AbsoluteThickness[2],
      Line[{{0, -11, 0}, {0, 13, 0}}]}],
   DisplayFunction -> $DisplayFunction,
   PlotRange -> All, ViewPoint -> {-0.5, -2.4, 1.9}];
```

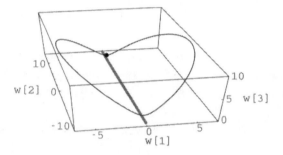

This image is quite informative! The rotation axis, represented by a point on the space curve, starts out at (0, 10, 1), just above the axis of symmetry, (0, 10, 0), but it travels very far from axis of symmetry before returning to it on the opposite side! The curve that is traced out by the changing instantaneous axis of rotation in the rotating frame of reference is referred to as the *polhode* curve. Note also that the trajectory looks like it will close up. Indeed, further computation confirms that the solution is periodic, and the discussion in section 17.2 of the ellipses defined by conservation laws provides a proof that periodicity holds. Here is a movie that shows how the instantaneous rotation axis moves along the curve. The printed image is a composite.

```
SpinTheBook[{0, 10, 1}, 3, RealWorld -> False, RotationAxis -> True,
    Polhode -> True, PlotRange -> {{-8, 8}, {-11, 11}, {-5, 11}},
    ViewPoint -> {0.8, 0.7, 0.4}]
```

You may also want to spin the book around the unstable axis yourself and see if the motion corresponds to the movie of the actual spinning shown below (we switch to the initial condition $(0, -10, -1)$ so that the movie corresponds to spinning the book holding it along the bottom). For example, try to catch the book so that the cover is face-up, but reads upside down compared to the way it is released. This occurs near the 1-second mark in the movie generated by the following code. The printed image shows only the first frame. Note that the curve traced out by the end of the angular velocity vector in the stationary frame (this curve is called the *herpolhode*) seems to lie in a plane. It does in fact lie in a plane, and this somewhat surprising fact will be discussed in section 17.4.

```
SpinTheBook[{0, -10, -1}, 2.4, Herpolhode -> True,
    RotationAxis -> True, Frames -> 50,
    PlotRange -> {{-6, 4.8}, {-10.5, 3.6}, {-6.1, 6}},
    ViewPoint -> {0.5, 1.4, 1.2}]
```

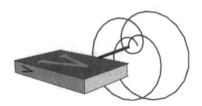

The images in the book's frame of reference were not too difficult to obtain: they required solving three fairly simple differential equations. But working only from these three equations, it would be not at all clear what the actual motion of the book in space looks like. In section 17.3 we will address the problem of getting images in the real world, which will require solving a system of twelve differential equations!

17.2 ■ Ellipsoids and the Conservation Laws

To better understand the solutions to Euler's equations, it helps tremendously to know that the equations admit two quadratic functions that must be conserved. Using the laws of conservation of energy and conservation of angular momentum, one can show that the following two quantities are preserved:

$$2E = I_1\omega_1^2 + I_2\omega_2^2 + I_3\omega_1^3$$

and

$$M^2 = I_1^2\omega_1^2 + I_2^2\omega_2^2 + I_3^2\omega_3^2$$

We can examine values of these expressions, using the solution we saved in the last section, to verify constancy.

```
Table[(Inertia /. Ivals) . (OMEGAsoln ^ 2), {t, 0, 1.8, 0.3}]
{30823., 30822.9, 30823.1, 30823.2, 30823.2, 30823.2, 30823.3}


Table[(Inertia^2 /. Ivals) . (OMEGAsoln ^ 2), {t, 0, 1.8, 0.3}]
{9.44003 10^6, 9.44 10^6, 9.44007 10^6, 9.44008 10^6,
    9.44008 10^6, 9.44008 10^6, 9.44012 10^6}
```

We see numerically that both quantities are conserved to several decimal places. Each of these conserved quantities defines an ellipsoid in **K**, and the solution therefore lies on the intersection of these two ellipsoids.

To see the ellipsoids and their intersection, we parametrize the ellipsoids by solutions to Euler's equations. The code is quite complicated as we use NDSolve and calculate the necessary polygons directly; the definitions of EllipseIntersection and PoinsotEllipsoid are in initialization cells in the version of this chapter on disk. The colored ellipsoids may be seen in color plate 25. The intersection of the two ellipsoids does not correspond to initial values that we have been using but were chosen to make clear that the two ellipsoids do intersect. For initial conditions too close to the second principal axis, one of the ellipsoids almost covers the other.

```
EllipseIntersection[ ];
```

In the above image, the light curves are solutions of Euler's equation all with the same energy levels and the dark curves are solutions all with same angular momentum. The thick black curves represent solutions that lie on the intersection of the two ellipsoids.

Much can be said of the motion of the body using the ellipsoids. The following image displays another view of the ellipsoid arising from conservation of energy, with the three axes of symmetry shown; this ellipsoid is called the *inertial ellipsoid*, or sometimes *Poinsot's ellipsoid*.

```
PoinsotEllipsoid[ ];
```

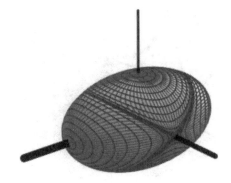

In the picture of the Poinsot ellipsoid, the three axes are represented in increasing thickness. The moments of inertia for the axes that we used were (420, 307, 123). So the first axis, represented by the thin vertical line, is the shortest and represents the greatest inertia. Spinning the ellipsoid itself around the first axis would require the most effort and spinning around the third axis (the fattest) would be easiest. The shape of the Poinsot ellipsoid thus resembles the shape of the body!

Directly from Euler's equations, we can see that a point on any of the principal axes corresponds to an equilibrium solution to the system. The ellipsoid above is parametrized by solutions to the equations. From the ellipsoid we can see that solutions for initial conditions near the first or third axes will remain close to the initial condition. It is easy to show, by linearizing the equations, that points on the first or third axes are stable equilibrium points.

But the two equilibrium points on the middle-length axis are unstable! Solutions jump back and forth between the positive and negative directions along the second principal axis. To illustrate what we mean by "jump back and forth" we look at a plot of the three Ω-components for an initial angular velocity of (0, 10, 1).

```
SystemSolutionPlot[OMEGAeqns /. Ivals, Evaluate[OMEGA],
    {t, 0, 2.4},
    InitialValues -> {0, 10, 1}, FastPlotting -> True,
    PlotStyle -> {{AbsoluteThickness[0.6]},
      {AbsoluteThickness[1.5]}, {AbsoluteThickness[3]}}];
```

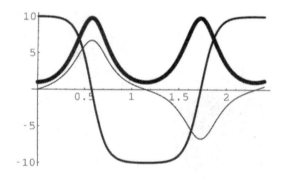

Note how the middle component of the angular velocity (medium-thick curve) spends most of its time near 10 and 10. That is, the vector stays near the ends of the unstable axis, quickly flipping from the vicinity of one end to the vicinity of the opposite end once it decides to move. One can see this behavior in the book-coordinate-system movie near the end of section 17.1.

It is often useful to know the exact value of the period. If a rough guess is known, then FindMinimum can be used to quickly zero in on the period, as follows.

```
normSq[v_] := N[v . v];
```

```
period = t /. Last[FindMinimum[
  normSq[OMEGAsoln - {0, 10, 1}], {t, {2.1, 2.4}}]]
```

```
2.30517
```

17.3 ■ Spinning a Book

In this section we develop the differential equations that allow us to show the actual spinning of the book in the real world. We would especially like to understand the motion near an unstable equilibrium. We hope you have already tossed a book and observed what happens. You could also try a tennis racket. The question then is whether the racket always flips over when it is grabbed by the handle and spun from a face-up (as opposed to the traditional face-vertical) starting position.

For the real-world view, we must work in the stationary coordinate system **k**. Everything we have done so far has been in the moving coordinate system of the body, **K**. Thus we must expend some energy to understand the function that takes **K** to **k**. More precisely, given a time t, we wish to understand the transformation B_t that takes the moving frame **K**, at time t, to the real-world frame **k**. Because both **k** and **K** are 3-dimensional coordinate systems, the function $B_t : \mathbf{K} \to \mathbf{k}$ corresponds to a rotation (because rotations are the only rigid motions of 3-space that fix the origin). Thus, for any fixed time t, there is a 3×3 rotation matrix B_t so that $\mathbf{q}_t = B_t\mathbf{Q}_t$ where **q** and **Q** are corresponding points in the two coordinate systems. Recall that a rotation matrix is an orthogonal matrix, which refers to a matrix whose columns are unit vectors, and with distinct columns being orthogonal (dot products are zero).

Temporarily writing B_t as $B(t)$ to emphasize the functional dependence on t, we start with the following equation [Arn, p. 131]: $B'(t)\mathbf{X} = B_t(\Omega \times \mathbf{X})$ for any vector **X** in our moving frame. Thus in terms of our coordinate system based on the book's axes, we get the following nine equations. First we define the nine B-variables we seek; then we define the nine equations.

```
B = Map[#[t] &, Array[b, {3, 3}], {2}]
```

```
{{b[1, 1][t], b[1, 2][t], b[1, 3][t]},
  {b[2, 1][t], b[2, 2][t], b[2, 3][t]},
  {b[3, 1][t], b[3, 2][t], b[3, 3][t]}}
```

```
(PosnEqns = Flatten[Thread /@
  Map[(D[B, t] . # == B . CrossProduct[OMEGA, #])&,
    IdentityMatrix[3]]]) //TableForm
```

```
(b[1, 1])'[t] == -(b[1, 3][t] w[2][t]) + b[1, 2][t] w[3][t]
(b[2, 1])'[t] == -(b[2, 3][t] w[2][t]) + b[2, 2][t] w[3][t]
(b[3, 1])'[t] == -(b[3, 3][t] w[2][t]) + b[3, 2][t] w[3][t]
(b[1, 2])'[t] == b[1, 3][t] w[1][t] - b[1, 1][t] w[3][t]
(b[2, 2])'[t] == b[2, 3][t] w[1][t] - b[2, 1][t] w[3][t]
(b[3, 2])'[t] == b[3, 3][t] w[1][t] - b[3, 1][t] w[3][t]
(b[1, 3])'[t] == -(b[1, 2][t] w[1][t]) + b[1, 1][t] w[2][t]
(b[2, 3])'[t] == -(b[2, 2][t] w[1][t]) + b[2, 1][t] w[2][t]
(b[3, 3])'[t] == -(b[3, 2][t] w[1][t]) + b[3, 1][t] w[2][t]
```

Notice that the equations for the matrix B also involve $w[1]$, $w[2]$, and $w[3]$. Thus, all told, we have a system of 12 differential equations for 12 unknown functions! This is a much more complicated situation than arose for any of the other models in this book. Using a coordinate system based on Euler angles and with an appropriate change of variables for the 3×3 system defined by Euler's equations, this system can be much simplified (see [Mur]). But *Mathematica* is able to handle the system we defined quite easily.

```
vars = Join[OMEGA, Flatten[B]]

{w[1][t], w[2][t], w[3][t], b[1, 1][t], b[1, 2][t], b[1, 3][t],
   b[2, 1][t], b[2, 2][t], b[2, 3][t], b[3, 1][t], b[3, 2][t], b[3, 3][t
```

Now we define `init` to set up the initial values corresponding to an initial spin. Then we use `NDSolve` to solve the resulting system of 12 equations; we store the first three functions of the solution in `ome`, and the last nine in `mat`.

```
init[{a_, b_, c_}] := Join[
  Thread[OMEGA == {a, b, c}],
  Thread[Flatten[B] == Flatten[IdentityMatrix[3]]]] /. t -> 0;

{ome, mat} = {OMEGA, B} /. First[NDSolve[
  Join[OMEGAeqns /. Ivals, PosnEqns, init[{0, 10, 1}]],
  vars, {t, 0, 3}]];
```

The preceding snippets of code are the key to generating a movie of a rotating book. The interested reader can examine the `SpinTheBook` code to see exactly how it is done. We would like to know the period, which we can get by minimizing the discrepancy from the start.

```
period = t /. Last[FindMinimum[normSq[ome - {0, 10, 1}],
  {t, {1.8, 2}}]]

2.30517
```

Let **M** be the angular momentum vector of the body relative to **0** in the moving frame. Then in terms of our principal axes coordinate system, **M** = Inertia · Ω. The angular momentum, **m**, in the stationary frame equals B_t**M**, and must remain fixed [Arn, p. 143].

```
Table[mat . ((Inertia /. Ivals) ome), {t, 0, 3, 0.5}]
{{0., 3070., 123.}, {0.0311852, 3070.01, 123.013},
    {0.00915364, 3070., 123.01}, {0.00286658, 3070., 123.014},
    {0.0138458, 3070., 123.018}, {0.00983978, 3070., 123.017},
    {0.00484086, 3070., 122.995}}
```

Having the transformation B_t from the moving coordinate system **K**, where we have Euler's equations describing the motion, to the stationary coordinate system **k**, we can actually write the code to "spin the book" which as we mentioned is done in the initialization cell at the beginning of the chapter on disk. After the discussion of the ellipsoids it is instructive to revisit the movie of the spinning book. The printed image is a composite from the movie with up-down and left-right motion added.

```
SpinTheBook[{0, -10, -1}, 2.4, Polhode -> True,
    TimeLabels -> False, RotationAxis -> True,
    PlotRange -> {{-6, 4.8}, {-10.5, 3.6}, {-6.1, 6}},
    ViewPoint -> {0.5, 1.4, 1.2}]
```

The book does indeed flip during the motion so that after one somersault it is spinning in nearly the opposite direction. But this is exactly what the pictures of the Ω-orbit predicted. The book spins near the initial value until some point when it jumps to a point on the opposite side of the ellipsoid, which corresponds to spinning around nearly the same axis but in the opposite direction.

An important point to make is that although Ω is periodic, the spinning of the book is not: after one period of Ω the book does not return to its initial position. The motion of the book over the next period of Ω will be similar to the first period, but from a different starting position. The image that follows show the curve corresponding to the real-world position of one corner of the book.

```
{ome, mat} = {OMEGA, B} /.  First[NDSolve[
    Join[OMEGAeqns /. Ivals, PosnEqns, init[{0, 10, 1}]],
    vars, {t, 0, 10}, MaxSteps -> 1000]];

ParametricPlot3D[Evaluate[mat . {0, 0, 4}], {t, 0, 10},
    PlotPoints -> 400, BoxRatios -> {1, 1, 1}];
```

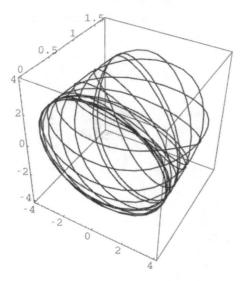

17.4 ■ Rolling the Poinsot Ellipsoid

We end with a graphic illustration of one of the most beautiful theorems of classical mechanics; it adds a great deal of insight into the motion of a rigid body that we have been considering, as it brings together many of the qualitative aspects of the motion that we have already covered. Recall that the instantaneous angular velocity vector Ω traces out a curve (the polhode) in the moving frame; and also the angular velocity vector ω traces out a curve in the stationary coordinate system as the book turns in the real world. This curve is called the herpolhode; it turns out that the herpolhode lies in a plane that is perpendicular to the angular momentum vector. (Proof: Working in the moving frame first, conservation of energy says that $I_1\omega_1^2 + I_2\omega_2^2 + I_3\omega_3^2$ is constant. But the angular momentum vector, $\mathbf{M}$, is $(I_1\omega_1, I_2\omega_2, I_3\omega_3)$. This means that $\mathbf{M}.\Omega$ is constant, and so the angular velocity lies in a plane normal to $\mathbf{M}$. But the transformation B_t is orthogonal, and so preserves dot products. Therefore $B_t(\mathbf{M}) \cdot B_t(\omega)$, which equals

$\mathbf{m} \cdot \omega$, is constant too, and ω lies in a plane normal to $\mathbf{m}$, as claimed.) This plane is called the *invariable plane* of the motion.

The conservation of energy ellipsoid, defined by $I_1\omega_1^2 + I_2\omega_2^2 + I_3\omega_3^2$, is called the inertia ellipsoid, or Poinsot's ellipsoid. It turns out that if we apply the transformation B to this ellipsoid, then the transformed ellipsoid is always tangent to the invariable plane and the center of the ellipsoid remains a fixed distance from the plane [Arn, p. 145]. Poinsot's theorem states that the rotation of the book corresponds to a rolling of the Poinsot ellipsoid without slipping along the invariable plane, so that the point of contact corresponds to the angular velocity vector. That is, the rolling takes place along the solution curve for Ω traced out on the ellipsoid. The rolling takes place without slippage because the point of contact is always along the instantaneous axis of rotation.

Here is an animation of the rolling Poinsot ellipsoid. The angular momentum, m, which remains fixed, is drawn as a thick light gray line. We also draw the angular velocity as a thin black line along with a thick blue curve for the herpolhode curve traced out by the angular velocity in the invariable plane. We extend the angular velocity below the invariable plane along with the herpolhode so that we can better see the herpolhode being traced out by the angular velocity. The polhode is the thick red curve that lies on the ellipsoid. The ellipsoid is shaded differently top and bottom to make it easier to see the orientation of the ellipsoid as it flips over.

```
RollPoinsot[{0, 10, 1}, 6];
```

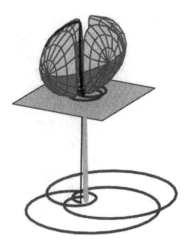

The closer we start to the second principal axis, the more the ellipsoid spins before rolling over. In the herpolhode the spinning corresponds to small loops near the center, that is, the point where the angular momentum hits the invariable plane. When the ellipsoid finally rolls over, the herpolhode makes a big loop before returning near to the center. It is possible (but difficult) to spin a book about its medium-length axis so that the cover faces up on the first revolution but down on the second revolution.

Watching the movie carefully, we see that after going through one period of the polhode, the ellipsoid has not returned to its original position but has turned through some angle. If this angle is not a rational multiple of 2π, then the herpolhode will never close up and over time the herpolhode will fill up some region of the plane.

Here we take a closer look at the herpolhode for an initial condition close to the second principal axis. In the code we calculate the intersection of the angular momentum vector with the invariable plane and use this to define two orthonormal unit vectors e1 and e2 in the invariable plane.

```
OMEGAinit = {0, -10, -1};
realBookSoln = First[NDSolve[
    Join[OMEGAeqns /. Ivals, PosnEqns, init[OMEGAinit]],
    vars, {t, 0, 15}, MaxSteps -> 5000]];
{ome, mat} = {OMEGA, B} /. realBookSoln;

AngMom = ((Inertia /. Ivals) OMEGAinit);
AngMomUnit = AngMom/Sqrt[normSq[AngMom]];
AngMomInt = AngMomUnit * normSq[(Sqrt[(Inertia /. Ivals)] OMEGAinit)]
            Sqrt[normSq[AngMom]];

e2 = (OMEGAinit - AngMomInt) / Sqrt[normSq[OMEGAinit - AngMomInt]];
e3 = CrossProduct[e2, AngMomUnit];

Show[Graphics[Line[Table[
  {(mat . ome - AngMomInt) . e2, (mat.ome - AngMomInt) . e3},
    {t, 0, 15, 0.01}]]], PlotRange -> All, Axes -> True,
    AspectRatio -> 1, AxesOrigin -> {-6, -6}];
```

Appendix

Usage Messages

AuxiliaryVariables is an option to ToSystem that specifies which auxiliary variables to use. It can be a single symbol (for single equations) or a list (for systems).

Color is an option to ProjectionPlot3D that, when False, causes a black and white image to appear.

ColorParametricPlot[f, {t, a, b}, (opts)] generates a parametric plot whose color changes continuously as a function of t. Options for ParametricPlot may be used. The Speed option allows the coloring to vary with the speed of motion instead.

ColorParametricPlot3D[f, {t, a, b}, (opts)] shows a 3-dimensional space curve using Hue to color the curves. The PlotStyle and PlotPoints options may be used.

ComputeWindow is an option to VisualDSolve and PhasePlot that works when StayInWindow is True. If set to Automatic, then computation shuts off at the viewing window, but a setting of {x0, x1} (VisualDSolve) or {{x0, x1}, {y0, y1}} (PhasePlot) causes that setting to be used instead.

DirectionArrow is an option to PhasePlot that, when True, adds arrows to show the direction of increasing t.

DirectionField is an option to VisualDSolve that when True, causes a field of slope lines to appear.

DirectionFieldPlot[f, {t, t0, t1, dt}, {x, x0, x1, dx}] gives an array of lines showing the slopes defined by the differential equation dx/dt = f. t0, t1, and dt give the range and step-size for the t-values, and similarly for the x-values. This is used by VisualDSolve and so the output is Graphics[] and not shown. All Graphics options may be used. The form DirectionFieldPlot[{fx, fy}, {x, x0, x1, dx}, {y, y0, y1, dy}] also works, and is used for autonomous systems {x' == fx(x, y), y' == fy(x, y)}. The lines will appear to have the same length (or, for systems, have length proportional to the vector), regardless of the aspect ratio used.

Equilibria[funcs, {x, xmin, xmax}, {y, ymin, ymax}, (opts)] returns the list of points where the funcs are 0. Options appropriate to FindRoot may be used. Also a table of eigenvalues corresponding to the equilibrium points is printed if ShowEigenvalues is True.

EquilibriaPlotPoints is an option to Equilibria that sets the PlotPoints for the contour plot that is used to generate seeds. To catch pairs of nearby equilibria it may be necessary to increase this.

EquilibriaPrecisionGoal is an option to Equilibria that eliminates duplicates with 10^prec.

EquilibriumPointStyle is an option to PhasePlot that sets the style of the equilibrium points (which are shown if ShowEquilibria is set to True). It can be a list or a single graphics primitive.

Euler is a choice for the Method option to VisualDSolve and PhasePlot. Number of steps is set by the EulerSteps option. One can't use special t-values; i.e., the initial values must all be for t0 equalling the tmin passed in {t, tmin, tmax}. Euler[] can be called separately and works in exactly the same way that NDSolve does, except that the starting t0 in {t, t0, t1} must appear in an initial condition as x[t0] == ***.

EulerSteps is an option to VisualDSolve and PhasePlot that controls the number of steps if Euler's method is selected.

FastPlotting is an option to VisualDSolve, SystemSolutionPlot, and Phase-Plot that, when True, causes the plot to be drawn just by using line segments through the interpolating points. This saves a lot of time over the adaptive plotting algorithm of Plot (since the interpolating points are generally nicely adaptive by construction); the curve will not be satisfactory in certain low-order cases, so we have set the default to False. Also be aware that if MaxSteps is large then the plots generated by this method might involve a huge number of data points. This option cannot be used with VisualDSolve if SymbolicSolution is set to True; and it works only if the default setting of ParametricPlotFunction is used. PlotPoints is ignored if FastPlotting is turned on.

FieldColor is an option to DirectionFieldPlot, PhasePlot, and PhaseLine that colors the flowing fish or vectors. Choices are: a single color or one of Hue, GrayLevel, RandomHue, or RandomGrayLevel. Hue and GrayLevel color the fish or vectors by relative lengths using hues or grays. FieldColor is also an option to PhaseLine, where it must be a single color.

FieldLength is an option to PhasePlot and DirectionFieldPlot that sets the length of the field lines or of the flowing fish. The default setting is 0.8, which uses 80% of the available space.

FieldLogScale is an option to PhasePlot, PhaseLine, and DirectionFieldPlot that scales the flowing fish or vectors proportionally to the longest fish. This is useful when a relatively few fish are several order of magnitudes longer than the fish in the interesting regions of the plot. FishLogScale -> 10 scales all the fish to the same length. Default is 1, which causes no change to the relative sizes. A typical setting in a troublesome situation might be 2, 3, or 4.

FieldMeshSize is an option to VisualDSolve and related functions that controls the mesh size of the direction field. It can be a single integer or a pair of integers.

FieldThickness is an option to PhasePlot, PhaseLine, and DirectionField-Plot that sets the thickness of the field lines.

FindMinMethod is an internal function and option value for the case that equilibria are sought for a nondifferentiable function.

FlowColor is an option to PhasePlot and FlowParametricPlot that sets the color of the fish shapes.

FlowField is an option to DirectionFieldPlot, PhasePlot, and PhaseLine. It replaces the traditional arrows by curvy fish shapes, randomly placed.

FlowParametricPlot[f, {t, a, b}, {x, xmin, xmax}, {y, ymin, ymax}] generates a parametric plot whose thickness increases with time to indicate direction of flow and speed of flow.

FlowThickness is an option to PhasePlot, DirectionFieldPlot and Phase-Line that, when fish are used, scales the head size, and so the overall thickness, of the fish. Default is 1.

FreehandAttempt[eqn, {t, tmin, tmax}, {x, xmin, xmax}, drawPts, (opts)] compares the curve determined by drawPts to the actual solution (in red) generated from the first points in drawPts. It works only for VisualDSolve examples, and passes its options to VisualDSolve.

GetPts[interFcn[t]] returns the set of y-interpolating points in the InterpolatingFunction interFcn, when viewed as a function y of t.

GetPtsFull[interFcn[t]] returns the set of interpolating points (t and y) in the InterpolatingFunction interFcn, when viewed as a function y of t.

GrayLevel[level] is a graphics directive that specifies the gray-level intensity with which graphical objects that follow should be displayed. GrayLevel may also be used as a setting to FieldColor, an option to PhasePlot.

GrayShading is an option to ColorParametricPlot that, when set to True, causes the colors to be replaced by shades of gray.

Grid[n] is a possible setting of the InitialValues option for VisualDSolve and causes an n-by-n grid to be used.

Hue[h] is a graphics directive which specifies that graphical objects which follow are to be displayed, if possible, in a color corresponding to hue h. Hue[h, s, b] specifies colors in terms of hue, saturation and brightness. Hue may also be used as a setting to FieldColor, an option to PhasePlot.

InflectionCurve is an option to VisualDSolve that causes the inflection curves to be drawn.

InflectionStyle is an option to VisualDSolve that specifies the style of the inflection lines.

InitialPointStyle is an option to VisualDSolve and related functions that specifies the style of the initial-value points. It can be a list or a single graphics primitive.

InitialValues is an option to VisualDSolve and related functions that sets the initial values. To specify one initial value, use {t0, x0} (for systems, use {t0, x0, y0, ...}). To specify several, use {{t0, x0}, {t1, x1}, ...}. However, one can also attach different t-ranges for efficiency (this does not work with SystemSolutionPlot): for one curve use {t0, x0, {tmin, tmax}} (or, for a PhasePlot of a system, {t0, x0, y0, (z0,...), {tmin, tmax}}); for several initial values, use several lists having this form. In the case of systems the t-value can be omitted, and it will be set to the lower bound in the main t-iterator. The possibility of varying the t-range makes NDSolve more efficient, since the ranges can be tweaked until they cause solutions that exist only in the final window; but StayInWindow->True accomplishes this too. Also, in VisualDSolve a setting of Grid[n] is a possibility for InitialValues.

IsoclinePlotPoints is an option to VisualDSolve that sets PlotPoints for the contour functions called for isoclines or inflection curves.

Isoclines is an option to VisualDSolve that causes positions of constant slope to be joined.

IsoclineShading is an option to VisualDSolve that shows a background shaded in gray according to the slopes. In its default form there are just two shades of gray, since the default setting for IsoclineValues is just $\{0\}$. This will generally be good enough, though you can get more regions by specifying Contours to be several values, or an integer, in which case a uniform set of levels is used.

IsoclineStyle is an option to VisualDSolve that specifies the style of the isoclines that are specified in IsoclineValues.

IsoclineValues is an option to VisualDSolve that specifies which slopes are to have isoclines. Default is $\{0\}$. If it is an integer, then that many equal-spaced contours are drawn.

LastPoint is a variable name that is used to store the last point(s) of a solution.

Logarithmic is a value to the PlotType option that requests a logarithmic (base 10) plot of the error.

MainThickness is an option to ProjectionPlot3D that sets the thickness, as a percentage of image width, of the main spacecurve.

Method is an option to Solve, related functions, and various numerical functions, which specifies what algorithm to use in evaluating the result. When used with VisualDSolve and PhasePlot it can force certain elementary methods to be used. Choices are: Euler and RungeKutta4. One cannot use these specific methods if StayInWindow is set to True.

MinorThickness is an option to ProjectionPlot3D that sets the thickness for the projected curves, as a percentage of image width.

NullclinePlot[$\{f1, f2\}$, $\{x, xmin, xmax\}$, $\{y, ymin, ymax\}$] produces, according to the setting of its options, the graphs of f1 = 0 and f2 = 0, a gray background that indicates the four possibilities for $\{Sign[f1], Sign[f2]\}$, and dots on the simultaneous roots (the equilibrium points). It works only on orbits for autonomous systems of two equations.

NullclinePlotPoints is an option to NullclinePlot that controls the resolution of the contour lines.

Nullclines is an option to NullclinePlot and PhasePlot that causes the nullclines to be shown.

NullclineShading is an option to NullclinePlot and PhasePlot that causes a coded gray background to show.

NullclineShadingMethod is an option to NullclinePlot that controls the algorithm used to shade the regions. Possibilities are Automatic or SumOf-Signs.

NullclineStyle is an option to NullclinePlot that sets the styles of the two nullclines. Thus it must be set to a pair, the first entry being a list of graphics primitives for the x-nullcline, and the second a list for the y-nullcline.

NumberFish is an option to PhaseLine and FlowParametricPlot that controls the number of fish.

ParametricPlotFunction is an option to PhasePlot that, if set to Color-ParametricPlot, causes a routine called ColorParametricPlot to be used to form the orbit(s). This colors the curve as the parameter increases. Another possible setting is FlowParametricPlot, which causes the Flow-ParametricPlot function to draw the orbit.

PhaseLine[eqn, x[t], {x, xmin, xmax}, (opts)] shows a one-dimensional phase portrait of a first-order, autonomous differential equation.

PhasePlot[eqn, unks, {t, tmin, tmax}, {x, xmin, xmax}, {y, ymin, ymax}, (opts)] shows the x-y orbit of the system eqn, which can be a system of 2 or more equations (not necessarily autonomous). Any options of ParametricPlot, Graphics, or NDSolve may be used; and there are LOTS of special options as well. Initial values must be specified by the InitialValues option. If there are 3 or more equations and the x-y iterators are replaced by 3 iterators, then the orbits are given as space curves, and options appropriate for 3D graphics may be used. In the 3D case, if the default ParametricPlotFunction is used, then PlotPoints will be set to 100, and can be changed. The plot is simply a uniform sequence of 100 line segments. If only a vector field or nullcline plot is wanted, the t-iterator, which is then irrelevant, may be omitted.

PlaneResolution is an option to ProjectionPlot3D that sets the grid of the three projection planes.

PlotLabels is an option to SystemSolutionPlot that is used when more than one plot is generated and accepts a list of plotlabels, one for each plot.

PlotRanges is an option to SystemSolutionPlot that is used when more than one plot is generated. It must be a list of plot ranges, one for each plot.

PlotStyle is an option for Plot, ParametricPlot, ListPlot, VisualDSolve, PhasePlot, and SystemSolutionPlot. PlotStyle -> style specifies that all lines or points are to be generated with the specified graphics directive, or list of graphics directives. PlotStyle -> {{style1}, {style2}, ...} specifies that successive lines generated should use graphics directives style1, style2, ...

PlotType is an option to ResidualPlot that allows a Logarithmic plot to be specified.

PlotVariables is an option to SystemSolutionPlot that specifies which variables are to be plotted.

PoincareSection[{eq1, eq2}, {t, tmin, tmax}, {x, xmin, xmax}, {y, ymin, ymax}, {x0, y0}, (opts)] returns a Poincare section of a system of two first-order ODEs, with the time interval determined by the TimeInterval option. {x0, y0} are the initial values. Equations are entered using x'[t], x[t], and so on. It also works on a single second-order equation via: **PoincareSection[eqn, {t, tmin, tmax}, {x, xmin, xmax}, {x', ymin, ymax}, {x0, xp0}]**.

PointStyle is an option to PoincareSection that sets the style in which the points are shown.

ProjectionPlot3D[{f, g}, {t, tmin, tmax}, {xmin, xmax}, {ymin, ymax}] gen-erates the space curve {f, g, t} together with its three planar projec-tions.

Rainbow is an option to VisualDSolve, FlowParametricPlot, PhasePlot, and SystemSolutionPlot that causes the curves corresponding to different ini-tial values to appear in a variety of hues. It also works with Poincare-Section, in which case different hues are used for the points.

RandomGrayLevel is a possible value to the FieldColor option for Direc-tionFieldPlot and PhasePlot which colors the fish or vectors with random shades of gray.

RandomHue is a possible value to the FieldColor option for DirectionField-Plot and PhasePlot which colors the fish or vectors with random hues.

ResidualPlot[eqn, soln, unk, {t, tmin, tmax}, (opts)] plugs a potential solution of a single (first- or second-order) ODE into the equation, subtracts, and plots the difference. unk is an expression representing the dependent function of t, typically x[t].

RKSteps is an option to RungeKutta4 and routines that call it (VisualD-Solve, PhasePlot, SecondOrderPlot) that controls the number of steps if the Runge-Kutta method is selected.

RungeKutta4 is a choice of method for VisualDSolve and PhasePlot, and forces the standard 4th-order Runge-Kutta method. Steps are set by the RKSteps option. RungeKutta4 can be called as a separate function; it works in exactly the same way that NDSolve does, except that the starting t0 in {t, t0, t1} must appear in an initial condition as x[t0] == ***.

SaveLastPoint is an option to VisualDSolve, PhasePlot, and SystemSolu-tionPlot that saves the final point(s) in LastPoint.

SavePoints is an option to PoincareSection that causes the points to be returned.

SaveSolution is an option to VisualDSolve and related functions that stores the solutions in a variable called 'Solution' so that it can be accessed after the use of VisualDSolve. It does not work if symbolic so-lutions are sought. In the case of PhasePlot, only the plotted solution are saved; i.e., if only two variables out of, say, 4, are plotted, only the plotted ones are saved. The saved object is an expression in t, not a pure function; to evaluate it at t0, use Solution /. t -> t0.

SecondOrderPlot[eqn, unks, {t, tmin, tmax}, (opts)] produces a plot of the solution to eqn, a second-order equation or system of such. Plotting variables are specified by the PlotVariables option, and x' may be used. This calls SystemSolutionPlot, and so plots may be requested in any of the forms that that routine produces. The default is to show graphs of all the functions and their derivatives on a single plot. If exactly two iterators such as {x, 0, 1}, {x', 0, 1} are included before the options, then PhasePlot is called and a phase plot is shown; options for PhasePlot may be used. The initial values must be given for the main variables first (unks), then the implicit auxiliary variables.

SeedsOnly is an option to Equilibria that returns rough values of the points. It is faster and often good enough for visualization. The main point is that this allows the nullclines to be split up into pieces at the equilibrium points.

Segments is an option to FlowParametricPlot that sets the number of segments used to draw each fish.

ShowEigenvalues is an option to Equilibria that causes a table of eigenvalues to be displayed.

ShowEquilibria is an option to PhasePlot and NullclinePlot that causes the equilibrium points to be shown.

ShowInitialValues is an option to VisualDSolve and related functions that when set to True, causes the initial-value points to be superimposed on the image.

ShowNumberOfSteps is an option to VisualDSolve, PhasePlot, and SystemSolutionPlot that yields a printed message with the number of steps that the numerical solver used (same as the number of interpolating points in the interpolating function representing the solution).

SolutionName is an option to VisualDSolve and related functions that accepts a string (!), which will become a global variable used to store the solution if SaveSolution is True.

Speed is an option to ColorParametricPlot that, when True, causes the coloring to run from blue (slowest) to red (fastest) corresponding to the speed of the particle. If GrayShading and Speed are both True, then the shading goes from black (slowest) to light gray (fastest).

StayInWindow is an option to PhasePlot and VisualDSolve that, when True, causes the plots to stop when they hit the window frame (or the max value in the t-iterator, or the setting of ComputeWindow). It also causes the orbits to expand in both directions from the initial point(s). When this setting is used, Method must be set to the default of Automatic. When StayInWindow is False, then the specified t-values are used. If any initial values come with specific t-limits, then they override a StayInWindow request. This allows specific t-limits to be used for periodic solutions that will never leave the window. The exit data, in a form suitable for later use with InitialValues, is returned as output.

SumOfSigns is a value to the NullclineShadingMethod that forces a different, simpler, but less highly resolved, method to be sued for shading nullclines.

SymbolicSolution is an option to VisualDSolve that requests the solver to precede the numerical approach with an attempt at finding a symbolic solution.

SystemSolutionPlot[equations, unknowns, {t, tmin, tmax}, (opts)] generates plots of x vs. t for every x that is a plotting variable (controlled by the PlotVariables option. Initial values must be included, and can have the forms {t0, x0, y0, ...} or just {x0, y0, ...} or a list of such; missing t-values are assumed to be tmin.

Thinning is an option to FreehandAttempt that thins out the user's data points to every third point, to speed up the interpolation to get the user's function.

TickLabelsOnly is an option to PhasePlot that causes the tick labels to appear, but no tick lines. This is useful because in the case of a shaded nullcline plot we have arranged for the tick lines to be on the outside of the axis. If the user does not want this, then this option allows him or her to turn it off, but retain the tick labels.

TimeInterval is an option to PoincareSection that sets the time interval in the section. The default setting of Automatic causes the interval to be set to 1/100 of the t-interval.

TimeScale is an option to PhaseLine that sets the time period that is used by NDSolve to calculate the flow from each initial condition. The initial conditions are chosen internally. Specifying a larger or smaller TimeScale affects how long each arrow or fish will be. The time-length of each fish will be TimeScale/NumberFish.

ToSystem[eqn, unk, t] turns a second-order DE to a system of two first-order DEs. If eqn is a list of second-order DEs and unk is a list of unknown functions, then a system of 2n equations is returned.

VectorField is an option to PhasePlot (autonomous case only) that, when True, causes a field of vectors to appear. Fish shapes are possible for phase plane plots via the FlowField option.

VisualDSolve[equation, {t, tmin, tmax}, {x, xmin, xmax}, (opts)] creates an image that shows the solutions to a first-order ODE superimposed on a field of slope lines. There are LOTS of options! And all of the options for Graphics, Plot, ContourPlot, and NDSolve may be used as well. The equation must be input in the standard form such as x'[t] + x[t] == t + x[t]^2. Note also that the first iterator defines the independent variable and the other iterator defines the dependent variable.

WindowShade is an option to VisualDSolve, PhasePlot, and PoincareSection that, if set to a color or gray shade, places a shaded rectangle in the viewing window.

References

[AK]   D. Armbruster and E. Kostelich. *Introductory Differential Equations: From Lin-earity to Chaos*. Addison-Wesley, Reading, Mass., 1996.

[Arn]   V. I. Arnold. *Mathematical Methods of Classical Mechanics*. Springer-Verlag, New York, 1989.

[BBS]   E. Batschelet, L. Brand, and S. Steiner. On the kinetics of lead in the human body. *Journal of Mathematical Biology*, 8:15–23, 1979.

[BCB]   R. L. Borrelli, C. Coleman, and W. E. Boyce. *Differential Equations Laboratory Workbook*. Wiley, New York, 1992.

[Cor]   R. M. Corless. *Error backward*, volume 172 of *Chaotic Numerics, Contemporary Mathematics*. American Mathematical Society, 1994. Eds., P. Kloeden, K. J. Palmer.

[deM]   N. de Mestre. The mathematics of projectiles in sport. *Australian Mathematical Society Lecture Series*, 6, 1990.

[Fro]   C. Frohlich. Aerodynamic effects on discus flight. *American Journal of Physics*, 49(12):1125–1132, December 1981.

[GH]   J. Guckenheimer and P. Holmes. *Nonlinear Oscillations, Dynamical Systems, and Bifurcations of Vector Fields*. Springer-Verlag, New York, 1983.

[Hub]   J. H. Hubbard. What it means to understand a differential equation. *College Mathematics Journal*, 25:372–384, 1994.

[Hub1]   J. H. Hubbard. The forced damped pendulum: chaos, complication and control. *CODEE newsletter*, pages 3–11, Spring 1995.

[HW]   J. H. Hubbard and B. H. West. *Differential Equations: A Dynamical Systems Approach*. Springer-Verlag, New York, 1991.

[KG]   D. Kaplan and L. Glass. *Understanding Nonlinear Dynamics*. Springer-Verlag, New York, 1995.

[KW]   R. Knapp and S. Wagon. Orbits worth betting on. *CODEE newsletter*, Spring 1996.

[KY1]   J. Kennedy and James A. Yorke. Basins of Wada. *Physica D*, 51:213–225, 1991.

[KY2]   J. Kennedy and James A. Yorke. The forced, damped, pendulum and the Wada property. *Continuum Theory and Dynamical Systems*, 149:157–181, 1993.

[Mur]   H. Murrell. Animation of rotating rigid bodies. *The Mathematica Journal*, 2:61–65, 1992.

[NY1]   Helena E. Nusse and James A. Yorke. Wada basin boundaries and basin cells. *Physica D*, 90:242–261, 1996.

[NY2]   Helena E. Nusse and James A. Yorke. Basins of attraction. *Science*, 271:1376–1380, March 8 1996.

[NY3]   Helena E. Nusse and James A. Yorke. The structure of basins of attraction and their trapping regions. *Ergodic Theory and Dynamical Systems*, Forthcoming.

[PR]   R. Pearl and L. J. Reed. On the rate of growth of the population of the United States since 1790 and its mathematical representation. *Proceedings of the National Academy of Sciences*, 6:275–288, 1920.

[Rei]   L. E. Reichl. *The Transition to Chaos*. Springer-Verlag, New York, 1992.

[RWK]   M. B. Rabinowitz, G. W. Wetherill, and J. D. Kopple. Lead metabolism in the normal human. *Science*, 182:725–727, 1973.

[SB]   C. Smith and N. Blachman. *The Mathematica Graphics Guidebook*. Addison-Wesley, Reading, Mass., 1995.

[Str]   S. H. Strogatz. *Nonlinear Dynamics and Chaos*. Addison-Wesley, Reading, Mass., 1995.

[SW]   D. Schwalbe and S. Wagon. Nullclines and equilibria, fish and balloons. *Mathematica in Education and Research*, 4(4):50–55, 1995.

[Wag]   S. Wagon. *Mathematica in Action*. W. H. Freeman, New York, 1991.

[Wag1]   S. Wagon. Getting inside plots. *Mathematica in Education and Research*, 3(4):43–46, 1994.

[Wol]   J. Wolkowiski. Exploring the double pendulum as a dynamical system. In *First International Mathematica Symposium*, pages 377–388. Computional Mechanics Publications, 1995.

Index

Index of VisualDSolve Functions

Index of Mathematica Objects

Subject Index